Principles of Nanotechnology

Principles of Nanotechnology

Timothy Sawyer

MURPHY & MOORE
www.murphy-moorepublishing.com

Murphy & Moore Publishing,
1 Rockefeller Plaza,
New York City, NY 10020, USA

ISBN: 978-1-63987-458-3

Cataloging-in-Publication Data

Principles of nanotechnology / Timothy Sawyer.
p. cm.
Includes bibliographical references and index.
ISBN 978-1-63987-458-3
1. Nanotechnology. 2. Nanostructured materials. 3. High technology. I. Sawyer, Timothy.
T174.7 .P75 2022
620.5--dc23

For information on all Murphy & Moore Publications
visit our website at www.murphy-moorepublishing.com

MURPHY & MOORE

Table of Contents

Preface

It is with great pleasure that I present this book. It has been carefully written after numerous discussions with my peers and other practitioners of the field. I would like to take this opportunity to thank my family and friends who have been extremely supporting at every step in my life.

The interdisciplinary field which focuses on the manipulation and manufacturing of materials and devices on atomic, molecular and supramolecular scale, is called nanotechnology. Nanomaterials are the materials which range between 1 to 100 nm. Molecular nanotechnology is the technology that is used for designing complex structures with the help of mechanosynthesis process. Bottom-up and top-down approaches are the two main approaches used in this field. Some top-down approaches include focused ion beams, solid-state techniques, etc. The bottom-up approaches include molecular-beam epitaxy, supramolecular approaches. Some of the theories where nanotechnology is applied are nanoelectronics, nanomedicine, nanopillars, biosensors, nanorods, drug delivery, etc. This book is a compilation of chapters that discuss the most vital concepts in the field of nanotechnology. It elucidates new techniques and their applications in a multidisciplinary approach. As this field is emerging at a rapid pace, the contents of this book will help the readers understand the modern concepts and applications of the subject.

The chapters below are organized to facilitate a comprehensive understanding of the subject:

Chapter - What is Nanotechnology?

Nanotechnology is the branch of science that deals with imaging, measuring and modeling of matter on an atomic, molecular and supermolecular scale. Nanostructure, nanosystems, nanoscience and nanoengineering, etc. are also studied within it. This chapter delves into the science of nanotechnology for an in-depth understanding of the subject.

Chapter - Branches of Nanotechnology

Nanotechnology is categorized into a varied number of branches. These include molecular nanotechnology, nanomechanics, nanoarchitectonics, nanoinformatics, nanometrology, nanofluidics, etc. This chapter has been carefully written to provide an easy understanding of the different branches of nanotechnology.

Chapter - Nanomaterials

The materials whose size lies between 1 and 100 nanometers are known as nanomaterials. Some of these are nanocomposites, silicon and carbon nanotubes, nanoshells, nanofibers, silicene, nanosheets and nanomesh. The topics elaborated in this chapter will help in gaining a better perspective of nanomaterials.

Chapter - Nanoelectronics

The utilization of nanotechnology in electronic components is referred to as nanoelectronics. Various components of nanoelectronics are nanoradios, nanomotors, nanoionic devices, nanobatteries, nanogenerators, nanosensors and nanowires. All these components associated with nanoelectronics have been carefully analyzed in this chapter.

Chapter - Applications

Nanotechnology has varied applications in the fields of food science, medicine, robotics, architecture, defense, space technology, industry, agriculture, etc. This chapter closely examines these applications of nanotechnology to provide an extensive understanding of the subject.

Chapter - Nanotechnology and the Environment

Environmental nanotechnology is the branch of nanotechnology that deals with the mitigation and prevention of damage to the environment. Some of its other divisions that aim at improving the environment are green nanotechnology, etc. This chapter discusses nanotechnology in relation with the environment in detail.

Timothy Sawyer

1

What is Nanotechnology?

Nanotechnology is the branch of science that deals with imaging, measuring and modeling of matter on an atomic, molecular and supermolecular scale. Nanostructure, nanosystems, nanoscience and nanoengineering, etc. are also studied within it. This chapter delves into the science of nanotechnology for an in-depth understanding of the subject.

NANOSCIENCE

Nanoscience is concerned with materials and systems whose structures and components exhibit novel and significantly improved physical, chemical and biological properties, phenomena and processes, because of their small nanoscale size. Structural features in the range of about 10 Å to 1000 Å, determine important changes as compared to the behavior of isolated molecules (10Å) or of bulk materials (>0.1 µm). Nanoscience aims to understand the novel properties and phenomena of nano-based entities.

What happens when the dimensions of structures are reduced to nanoscale levels?

Reducing the dimensions of structures leads to entities with novel properties, such as carbon nanotubes, quantum wires and dots, thin films, DNA based structures, and laser emitters. Such new forms of materials and devices herald a revolutionary age for science and technology, provided that we can discover and fully utilize the underlying principles.

Benefits of Studying the Behavior and Interactions of Nanoscale Materials

New behavior at the nanoscale is not necessarily predictable from that observed at large size scales. Important changes in behavior are caused not by the order of magnitude size reduction, but also by new phenomena such as size confinement, predominance of interfacial phenomena, quantum mechanics and Coulomb blockade. It is notable that all relevant phenomena at the nanoscale are caused by the tiny size of the organized structure as compared to molecular scale, and by the interactions at their predominant and complex interfaces.

Aims of Nanoscience

Once we are able to control feature size, we can enhance material properties and device functions beyond those that we currently know or even imagine. Nanotechnology aims to gain control of structures and devices at the atomic, molecular and supramolecular levels, and to learn how to efficiently manufacture and use these devices.

NANOENGINEERING

Nanoengineering is a branch of engineering that deals with all aspects of the design, building, and use of engines, machines, and structures on the nanoscale. At its core, nanoengineering deals with nanomaterials and how they interact to make useful materials, structures, devices and systems.

Nanoengineering is not exactly a new science, but, rather, an enabling technology with applications in most industries from electronics, to energy, medicine, and biotechnology.

While the term nanoengineering is often used synonymously with the more general term nanotechnology, the former technically focuses more closely on the engineering aspects of the field, as opposed to the broader science and general technology aspects that are encompassed by the latter.

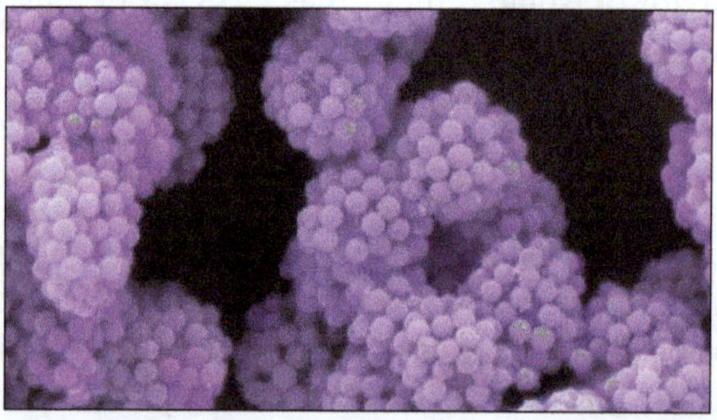

Tiny particles transformed into 'LEGO-like' modular building blocks.

In general, engineering is the branch of science and technology concerned with the design, building, and use of engines, machines, and structures. Correspondingly, but at the scale of atoms and molecules, nanoengineeering exploit the unique properties of nanoscale materials (size and quantum effects) in order to design and manufacture devices and systems that possess entirely new functionality and capabilities.

The properties of materials can be different at the nanoscale for two main reasons:

First, nanomaterials have a relatively larger surface area when compared to the same mass of material produced in a larger form. This can make materials more chemically reactive (in some cases materials that are inert in their larger form are reactive when produced in their nanoscale form), and affect their strength or electrical properties.

Second, quantum effects can begin to dominate the behavior of matter at the nanoscale – particularly at the lower end – affecting the optical, electrical and magnetic behavior of materials. Materials can be produced that are nanoscale in one dimension (for example, nanowires, nanorods and nanotubes), in two dimensions (plate-like shapes like nanocoatings, nanolayers, and graphene) or in all three dimensions (for example, nanoparticles).

Nanoscale objects are difficult to manipulate, as they are too tiny to see directly by eye, far too small to hold, and often have incompatible surfaces for assembling into ordered structures. Therefore the fabrication of complex nanoarchitectures requires sophisticated techniques of nanoscale engineering. To do this, nanoengineers are employing a number of methods to leverage the manipulation of materials on an atomic and molecular scale for (ultimately) industrial purposes.

Among the many challenges that researchers have to overcome in developing nanoengineering techniques and processes, the requirement for extremely precise, nanometer-scale control of positioning and shaping of objects is one of the most vexing.

Nanopatterning and Nanostructuring of Surfaces

There are different ways of fabricating functional nanostructures that you hear most about: top-down and bottom-up methods.

Top-down nanofabrication predominantly is done by photolithography, which is the standard workhorse employed in today's semiconductor industry, and other top-down methods where you start by taking a block of material and remove the bits and pieces you don't want until you get the shape and size you do want.

In contrast to the deterministic nature of top-down processes, bottom-up processes are driven by a combination of thermodynamics and kinetics, which then determines the yield of the desired structure.

Bottom-up processes typically don't need expensive tooling to create nanoscale structures, and scaling to large volumes is potentially straightforward. With the application of the tools of chemical synthesis, quantum dots, plasmonically active particles, carbon nanotubes, metallic nanowires, and multifunctional particles for medical applications have been successfully produced in manufacturing quantities.

Given the vast, and still rising, cost of optical lithography tools, researchers have considered alternative patterning technologies such as electron beam lithography (EBL), and nanoimprint technology (NIL) in order to enable the manufacturing of next-generation integrated circuits, flash memory, and hard disk drives.

Now that the length scales attainable by top-down lithography are approaching that of bottom-up self-assembly found in polymers and small molecules, scientists are increasingly looking at bottom-up patterning technologies based on self-assembly.

The utilization of organic building blocks as self-assembled etch masks is particularly attractive since the lithographic information is contained in the self-assembling material itself, rather than provided in an exposure step.

Architecture is key when designing new materials and DNA is probably the most programmable biomaterial for creating a wide range of rationally designed and functionally enhanced nanostructures.

DNA nanotechnology employs DNA as a programmable building material for self-assembled, nanoscale structures with precisely controlled structures, that can lead to improved surface properties relevant to biosensing, materials science, and cell biology.

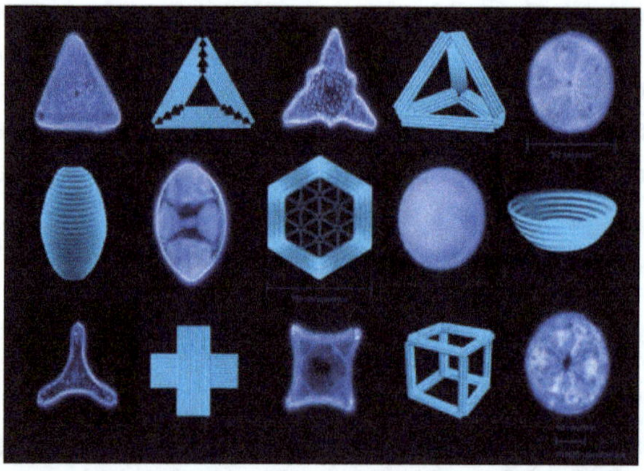

A selection of nanostructures built using DNA origami, alongside naturally-occurring diatoms – single-celled organisms that come in many beautiful and elaborate forms. They are ubiquitous inhabitants of the world's lakes, rivers, and oceans. A scale shows the sizes of the nanostructures and diatoms.

3D Printing

Fabrication of three-dimensional (3D) objects through direct deposition of functional materials – also called additive manufacturing – has been a subject of intense study in the area of macroscale manufacturing for several decades. These 3D printing techniques are reaching a stage where desired products and structures can be made independent of the complexity of their shapes – even bioprinting tissue is now in the realm of the possible.

Applying 3D printing concepts to nanotechnology could bring similar advantages to nanoengineering – speed, less waste, economic viability – than it is expected to bring to manufacturing technologies.

In addition, pre-patterned micro- or nanostructures could be used as substrates, allowing researchers to realize unprecedented manufacturing flexibility, functionality and complexity at the nanoscale.

Already, 3D printing can now be used to print lithium-ion microbatteries the size of a grain of sand.

This image shows the interlaced stack of electrodes that were printed layer by layer to create the working anode and cathode of a microbattery.

There are various approaches to 3D nanoprinting. In one, researchers use a novel nanoscale additive manufacturing technique termed Nanotribological Printing, which creates structures through tribomechanical and tribochemical surface interactions at the contact between a substrate and an atomic force microscope probe, where material pattern formation is driven by normal and shear contact stresses.

NANOTECHNOLOGY

Nanotechnology is the manipulation and manufacture of materials and devices on the scale of atoms or small groups of atoms. The "nanoscale" is typically measured in nanometres, or billionths of a metre (nanos, the Greek word for "dwarf," being the source of the prefix), and materials built at this scale often exhibit distinctive physical and chemical properties due to quantum mechanical effects. Although usable devices this small may be decades away (see microelectromechanical system), techniques for working at the nanoscale have become essential to electronic engineering, and nanoengineered materials have begun to appear in consumer products. For example, billions of microscopic "nanowhiskers," each about 10 nanometres in length, have been molecularly hooked onto natural and synthetic fibres to impart stain resistance to clothing and other fabrics; zinc oxide nanocrystals have been used to create invisible sunscreens that block ultraviolet light; and silver nanocrystals have been embedded in bandages to kill bacteria and prevent infection.

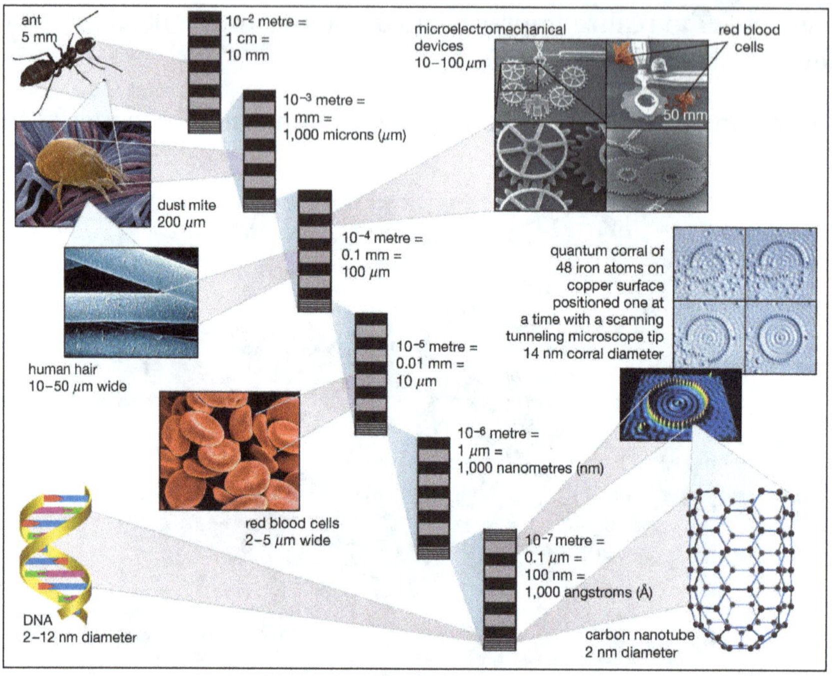

Examples from biological and mechanical realms illustrate various "orders of magnitude" (powers of 10), from 10^{-2} metre down to 10^{-7} metre.

Possibilities for the future are numerous. Nanotechnology may make it possible to manufacture lighter, stronger, and programmable materials that require less energy to produce than conventional materials, that produce less waste than with conventional manufacturing, and that promise greater fuel efficiency in land transportation, ships, aircraft, and space vehicles. Nanocoatings for both opaque and translucent surfaces may render them resistant to corrosion, scratches, and radiation. Nanoscale electronic, magnetic, and mechanical devices and systems with unprecedented levels of information processing may be fabricated, as may chemical, photochemical, and biological sensors for protection, health care, manufacturing, and the environment; new photoelectric materials that will enable the manufacture of cost-efficient solar-energy panels; and molecular-semiconductor hybrid devices that may become engines for the next revolution in the information age. The potential for improvements in health, safety, quality of life, and conservation of the environment are vast.

At the same time, significant challenges must be overcome for the benefits of nanotechnology to be realized. Scientists must learn how to manipulate and characterize individual atoms and small groups of atoms reliably. New and improved tools are needed to control the properties and structure of materials at the nanoscale; significant improvements in computer simulations of atomic and molecular structures are essential to the understanding of this realm. Next, new tools and approaches are needed for assembling atoms and molecules into nanoscale systems and for the further assembly of small systems into more-complex objects. Furthermore, nanotechnology products must provide not only improved performance but also lower cost. Finally, without integration of nanoscale objects with systems at the micro- and macroscale (that is, from millionths of a metre up to the millimetre scale), it will be very difficult to exploit many of the unique properties found at the nanoscale.

Nanotechnology is highly interdisciplinary, involving physics, chemistry, biology, materials science, and the full range of the engineering disciplines. The word nanotechnology is widely used as shorthand to refer to both the science and the technology of this emerging field. Narrowly defined, nanoscience concerns a basic understanding of physical, chemical, and biological properties on atomic and near-atomic scales. Nanotechnology, narrowly defined, employs controlled manipulation of these properties to create materials and functional systems with unique capabilities.

In contrast to recent engineering efforts, nature developed "nanotechnologies" over billions of years, employing enzymes and catalysts to organize with exquisite precision different kinds of atoms and molecules into complex microscopic structures that make life possible. These natural products are built with great efficiency and have impressive capabilities, such as the power to harvest solar energy, to convert minerals and water into living cells, to store and process massive amounts of data using large arrays of nerve cells, and to replicate perfectly billions of bits of information stored in molecules of deoxyribonucleic acid (DNA).

There are two principal reasons for qualitative differences in material behaviour at the nanoscale (traditionally defined as less than 100 nanometres). First, quantum mechanical effects come into play at very small dimensions and lead to new physics and chemistry. Second, a defining feature at the nanoscale is the very large surface-to-volume ratio of these structures. This means that no atom is very far from a surface or interface, and the behaviour of atoms at these higher-energy sites have a significant influence on the properties of the material. For example, the reactivity of a metal catalyst particle generally increases appreciably as its size is reduced—macroscopic gold is chemically inert, whereas at nanoscales gold becomes extremely reactive and catalytic and even melts at a lower temperature. Thus, at nanoscale dimensions material properties depend on and change with size, as well as composition and structure.

Using the processes of nanotechnology, basic industrial production may veer dramatically from the course followed by steel plants and chemical factories of the past. Raw materials will come from the atoms of abundant elements—carbon, hydrogen, and silicon—and these will be manipulated into precise configurations to create nanostructured materials that exhibit exactly the right properties for each particular application. For example, carbon atoms can be bonded together in a number of different geometries to create variously a fibre, a tube, a molecular coating, or a wire, all with the superior strength-to-weight ratio of another carbon material—diamond. Additionally, such material processing need not require smokestacks, power-hungry industrial machinery, or intensive human labour. Instead, it may be accomplished either by "growing" new structures through some combination of chemical catalysts and synthetic enzymes or by building them through new techniques based on patterning and self-assembly of nanoscale materials into useful predetermined designs. Nanotechnology ultimately may allow people to fabricate almost any type of material or product allowable under the laws of physics and chemistry. While such possibilities seem remote, even approaching nature's virtuosity in energy-efficient fabrication would be revolutionary.

Even more revolutionary would be the fabrication of nanoscale machines and devices for incorporation into micro- and macroscale systems. Once again, nature has led the way with the fabrication of both linear and rotary molecular motors. These biological machines carry out such tasks as muscle contraction (in organisms ranging from clams to humans) and shuttling little packets of material around within cells while being powered by the recyclable, energy-efficient fuel adenosine triphosphate. Scientists are only beginning to develop the tools to fabricate functioning systems at such small scales, with most advances based on electronic or magnetic information processing and storage systems. The energy-efficient, reconfigurable, and self-repairing aspects of biological systems are just becoming understood.

The potential impact of nanotechnology processes, machines, and products is expected to be far-reaching, affecting nearly every conceivable information technology, energy source, agricultural product, medical device, pharmaceutical, and material used in manufacturing. Meanwhile, the dimensions of electronic circuits on semiconductors continue to shrink, with minimum feature sizes now reaching the nanorealm, under 100 nanometres. Likewise, magnetic memory materials, which form the basis of hard disk drives, have achieved dramatically greater memory density as a result of nanoscale structuring to exploit new magnetic effects at nanodimensions. These latter two areas represent another major trend, the evolution of critical elements of microtechnology into the realm of nanotechnology to enhance performance. They are immense markets driven by the rapid advance of information technology.

Properties at the Nanoscale

At nanoscale dimensions the properties of materials no longer depend solely on composition and structure in the usual sense. Nanomaterials display new phenomena associated with quantized effects and with the preponderance of surfaces and interfaces.

Quantized effects arise in the nanometre regime because the overall dimensions of objects are comparable to the characteristic wavelength for fundamental excitations in materials. For example, electron wave functions in semiconductors are typically on the order of 10 to 100 nanometres. Such excitations include the wavelength of electrons, photons, phonons, and magnons, to name a few. These excitations carry the quanta of energy through materials and thus determine the dynamics of their propagation and transformation from one form to another. When the size of structures is comparable to the quanta themselves, it influences how these excitations move through and interact in the material. Small structures may limit flow, create wave interference effects, and otherwise bring into play quantum mechanical selection rules not apparent at larger dimensions.

Electronic and Photonic Behavior

Quantum mechanical properties for confinement of electrons in one dimension have long been exploited in solid-state electronics. Semiconductor devices are grown with thin layers of differing composition so that electrons (or "holes" in the case of missing electron charges) can be confined in specific regions of the structure (known as quantum wells). Thin layers with larger energy bandgaps can serve as barriers that restrict the flow of charges to certain conditions under which they can "tunnel" through these barriers—the basis of resonant tunneling diodes. Superlattices are periodic structures of repeating wells that set up a new set of selection rules which affect the conditions for charges to flow through the structure. Superlattices have been exploited in cascade lasers to achieve far infrared wavelengths. Modern telecommunications is based on semiconductor lasers that exploit the unique properties of quantum wells to achieve specific wavelengths and high efficiency.

The propagation of photons is altered dramatically when the size and periodicity of the transient structure approach the wavelength of visible light (400 to 800 nanometres). When photons propagate through a periodically varying dielectric constant—for example, semiconductor posts surrounded by air—quantum mechanical rules define and limit the propagation of the photons depending on their energy (wavelength). This new behaviour is analogous to the quantum mechanical rules that define the motion of electrons through crystals, giving bandgaps for semiconductors. In one dimension, compound semiconductor superlattices can be grown epitaxially with the alternating layers having different dielectric constants, thus providing highly reflective mirrors for specific wavelengths as determined by the repeat distance of layers in the superlattice. These structures are used to provide "built-in" mirrors for vertical-cavity surface-emitting lasers, which are used in communications applications. In two and three dimensions, periodic structures known as photonic crystals offer additional control over photon propagation.

Photonic crystals are being explored in a variety of materials and periodicities, such as two-dimensional hexagonal arrays of posts fabricated in compound semiconductors or stacked loglike arrays of silicon bars in three dimensions. The dimensions of these structures depend on the wavelength of light being propagated and are typically in the range of a few hundred nanometres for wavelengths in the visible and near infrared. Photonic crystal properties based on nanostructured materials offer the possibility of confining, steering, and separating light by wavelength on unprecedented small scales and of creating new devices such as lasers that require very low currents to initiate lasing (called near-thresholdless lasers). These structures are being extensively investigated as the tools for nanostructuring materials are steadily advancing. Researchers are particularly interested in the infrared wavelengths, where dimensional control is not as stringent as at the shorter visible wavelengths and where optical communications and chemical sensing provide motivation for potential new applications.

Magnetic, Mechanical and Chemical Behavior

Nanoscale materials also have size-dependent magnetic behaviour, mechanical properties, and chemical reactivity. At very small sizes (a few nanometres), magnetic nanoclusters have a single magnetic domain, and the strongly coupled magnetic spins on each atom combine to produce a particle with a single "giant" spin. For example, the giant spin of a ferromagnetic iron particle rotates freely at room temperature for diameters below about 16 nanometres, an effect termed superparamagnetism. Mechanical properties of nanostructured materials can reach exceptional strengths. As a specific example, the introduction of two-nanometre aluminum oxide precipitates into thin films of pure nickel results in yield strengths increasing from 0.15 to 5 gigapascals, which is more than twice that for a hard bearing steel. Another example of exceptional mechanical properties at the nanoscale is the carbon nanotube, which exhibits great strength and stiffness along its longitudinal axis.

The preponderance of surfaces is a major reason for the change in behaviour of materials at the nanoscale. Since up to half of all the atoms in nanoparticles are surface atoms, properties such as electrical transport are no longer determined by solid-state bulk phenomena. Likewise, the atoms in nanostructures have a higher average energy than atoms in larger structures, because of the large proportion of surface atoms. For example, catalytic materials have a greater chemical activity per atom of exposed surface as the catalyst is reduced in size at the nanoscale. Defects and impurities may be attracted to surfaces and interfaces, and interactions between particles at these small dimensions can depend on the structure and nature of chemical bonding at the surface. Molecular monolayers may be used to change or control surface properties and to mediate the interaction between nanoparticles.

Surfaces and their interactions with molecular structures are basic to all biology. The intersection of nanotechnology and biotechnology offers the possibility of achieving new functions and properties with nanostructured surfaces. In this surface- and interface-dominated regime, biology does an exquisite job of selectively controlling functions through a combination of structure and chemical forces. The transcription of information stored in genes and the selectivity of biochemical reactions based on chemical recognition of complex molecules are examples where interfaces play the key role in establishing nanoscale behaviour. Atomic forces and chemical bonds dominate at these dimensions, while macroscopic effects—such as convection, turbulence, and momentum (inertial forces)—are of little consequence.

NANOSTRUCTURE

Nanostructures have been described as 'novel materials whose size of elemental structure has been engineered at the nanometer scale'. Materials in the nanometer size range commonly exhibit fundamentally new behavior. Moreover, intervention in the properties of materials at the nanoscale enables the creation of materials and devices with enhanced or completely new characteristics and functionalities. Understanding the science of nanomaterials is important and curiosity driven not only because of the fascinating nature of the subject but also for overwhelming and novel applications of nanoscale systems in almost all branches of technology. Nanotechnology can be understood as a technology of design, fabrication and application of nanostructures and nanomaterials.The field of nanoscience and nanotechnology is interdisciplinary in nature and it is being perused by physicists, chemists, materials scientists, biologists, engineers, computer scientists etc. Nanomaterials may be classified on the basis of dimensionality and modulation. Some special nanostructures like nanotubes, nanoporous materials, aerogels, zeolites, core-shell structures have also come up with their novel characteristics. A number of methods have been used for the synthesis of nanostructure with various degrees of success and many direct and indirect techniques are employed for their characterization. The fact, which makes the nanostructures interesting, is that the properties become size dependent in nanometer range because of surface effect and quantum confinement effect.

The geometric structure, chemical bonds, ionization potential, electronic properties, optical properties, mechanical strength, thermal properties, magnetic properties etc. are all affected by particle sizes in nanometer range. Nanomaterials exhibit properties often superior to those of conventional coarse-grained materials. These include increased strength/hardness, enhanced diffusivity, improved ductility/toughness, reduced density, reduced elastic modulus, higher electrical resistivity, increased specific heat, higher thermal expansion coefficient, lower thermal conductivity, increased oscillator strength, blue shifted absorption, increased luminescence and superior soft magnetic properties in comparison to conventional bulk materials. All of these properties are being extensively investigated to explore possible applications. The fascinating field of nanotechnology has wide range of applications. Use of nanostructutred materials has produced transistors with record low speed and lasers with low threshold current. These are being used in compact disk player systems, low noise amplification in satellite receivers as sources for fiber optic communication etc. Constructive applications of nanomaterials include self-cleaning glass, UV resistant wood coating etc. Nanoscale devices are being used in medical field also for diagnosis, treatment and prevention of diseases and in drug delivery system, magnetic resonance imaging, radioactive tracers etc. Importance of nanotechnology is growing day by day. Many more applications may be possible with the novel and peculiar properties of nanostructures.

Nanostructures have been described as 'novel materials whose size of elemental structure has been engineered at the nanometer scale'. Dictated by the demands of miniaturization, increasing efforts are being made to synthesize, understand and apply the materials with reduced dimensions. The interest in nanomaterials is growing at a dramatic rate due to realization that reduced dimensions in nanometer regime can alter and improve the properties of materials. Nanomaterials have attracted widespread attention since 1990s because of their specific features that differ from bulk materials. This has opened up with tailor-made materials enormous possibilities to meet challenges in different disciplines like medicine, biotechnology, optoelectronoics, engineering etc. of new, effective and efficient devices, drugs or tools. It is expected that the emerging technology will focus on challenges of miniaturization and energy saving, covering different technologies that are existing today. It is therefore not very surprising to see that scientist from different disciplines are joining hands to contribute to this dynamic growing field of nanomaterials.

The origin of the term 'nano' comes from Greek word for 'dwarf' but in scientific jargon 'nano' means 10^{-9}. So a nanometer means a billionth of a meter, that is 10 times the diameter of hydrogen atom or approximately 1/10000 of the diameter of human hair. Nanotechnology can be understood as a technology of design, fabrication and application of nanostructures and nanomaterials. The concept of nanotechnology, however, dates back to the history of the famous lecture that Noble Laureate Richard Feynman gave in 1959: "The principles of physics, as far as I can see, do not speak against the possibility of maneuvering things atom by atom." Later Drexler expanded Feynman's philosophical thinking in a most stimulating manner in 1960, in his book: "Nanotechnology is the principle of manipulation atom by atom, through control of structure of matter at molecular level. It entails the ability to build molecular systems with atom by atom, yielding a variety of nanomechanics". Chemistry deals with the study of atoms and molecules, a realm of matter whose dimensions are generally less than 1 nm, while solid state physics deals with matter of essentially an infinite array of bound atoms or molecules of dimensions greater than 100 nm. A significant gap exists between the two regimes, which is mesoscopic or intermediate regime of nanostructures from 1 to 100 nm or approximately 10 to 10^6 atoms or molecules per particle. In this nanoscale regime neither quantum chemistry nor classical laws of solid state physics holds and the matter exhibits new size dependent properties. Such materials have come to be known as nanomaterials.

Changes in Properties

All the materials—may be metals, semiconductors or insulators—have size dependent physico-chemical properties below a certain critical size depending upon the details of the materials. For most of the materials, it is below 100nm. Thus geometrical structure, chemical bonds, ionization potential, electronic properties, optical properties, mechanical strength, melting point, magnetic properties etc. are all affected by the particle sizes:

- Mechanical properties: It has been observed that nanocrystalline materials have reduced elastic modulus and density by 30% or less. This may be due to large free volume at interfacial space and increased average inter-atomic distance. Hardness or strength increases by 4-5 times with decreasing size. Diffusivity of nanomaterials also increases up to double of the initial. Numerous interfaces provide paths for diffusion. Increased diffusivity increases sinterability at lower temperatures.

- Thermal properties: It is found that melting point is lowered with decreasing particle size. It may be reduced to half of the original. The specific heat and thermal expansion may increase up to 50% or more with reducing particle size.
- Optical properties: The color and transparency of materials change by changing the size of nanoparticles. Hence these can be controlled by processing parameters. Due to the increase in effective band gap of materials, absorption spectra shift towards higher energy and for small particle; it may show stepwise absorption by decreasing size of nanocrystals. Enhanced luminescence with fast response has been observed by decreasing size of nanocrystals; this is because of changes in electronic structure and useful in fast response devices with emission of desired color. Nanoparticles may be used in lasers also since these can operate at lower threshold. Confinement of photons and phonons in nanoparticles also affect the Raman spectra. Nonlinear optical properties are also observed in semiconductor clusters in glass or polymer matrix.
- Magnetic Properties: Saturation magnetization values of nanoparticles are smaller but coercive values are much larger than their polycrystalline counterparts because of their high surface to volume ratio and increased effective anisotrophy. Curie temperature of ferromagnetic materials decreases with decreasing size of nanoparticles, and hence the substance remains paramagnetic even below usual Curie temperature showing super paramagtetism. In nanocrystalline phase, each particle is a single ferromagnetic domain.
- Electrical properties: Electrical properties are also affected by the size in nanometer range. Electrical conductivity is reduced by decreasing the size of nanocrystals. Ferroelectric materials (e.g.PbT_iO_3) become non-ferroelectric at reduced sizes (20 nm). Transition temperature of superconducting materials (e.g.YBa_2C-u_3O_7) is reduced and these become nonsuperconducting for smaller sizes.

This has opened up a huge possibility of synthesising and manufacturing the materials with desired characteristics. It should be noted that for a particular material there is always a critical size below which properties are size dependent. In these materials the length of reduced dimension has to be smaller than the phase coherent length or the mean free path scattering length for electrons in the materials.

Reasons for Novel Properties

Nanomaterials exhibit properties often superior to those of conventional coarse-grained materials and in addition, the properties can be tailored by controlling the size. The changes in the properties of nanomaterials are driven mainly by three factors:

- Increase in surface area to volume ratio: Atoms and molecules at surface or interface have different environment and bonding configuration, therefore these exhibit different characteristics. In micron size particles, the fraction of surface atoms is less than 10-8 and so they cause little effect. As size is reduced, relative number of atoms on surface increases inversely as particle size. For small sizes of the order of few 100 nm, the fraction of surface atoms is large and therefore influences the properties by greater amount making the properties size dependent.
- Quantum size effect: When the size of the particle is comparable to phase coherent length of electrons, the energy spectrum is quantized into discrete levels. The effect is observable for particle sizes of the order of few 10 nm. In case of metals the continuous energy levels are discretized with energy spacing E_f /N, where E_f is Fermi energy and N is number of electrons present. The discretization in energy levels is experienced when their spacing exceeds the thermal energy. Quantum size effect is more pronounced in semiconductor nanoparticles because of moderate forbidden energy gap. Quantum size effect makes the energy states at the edges of valence and conduction bands forbidden, increasing the effective band gap, which affects the electronics and optical properties.
- Lattice contraction: At very small sizes of the order of a few nm, lattice parameters may be reduced because of inward interatomic forces. Structural phase change, such as from cubic to hexagonal, has also been observed in this size range.

Intervention in the properties of materials at the nanoscale enables the creation of materials and devices with enhanced or completely new characteristics and functionalities. Therefore these find wide applications in various devices and improved performance is achievable and with the help of these materials.certain new novel devices have also become possible.

Classification

Interestingly, the reduction in the physical size of the materials may be only in one direction, giving rise to a thin film or reduction may be in two dimension making it wire or reduction may be in all direction making it a point. This puts certain restrictions on the motion of charge carriers. It is not difficult to visualize that for a bulk (3-D) material charge carriers are free to move in all directions; whereas for a plainer material or thin film (2-D), charge carriers are confined only in a plane. The charge carries in a wire or 1-D material will be allowed to flow only in one direction; whereas charge carriers inside a cluster, nanoparticle or quantum dot (0-D) are confined in a very small region in space. This gives rise to quantum confinement and most of the properties like electrical conductivity, optical absorption, thermal conductivity etc. depend on the electronic structure and vary due to this effect. The effect occurs when the dimension is of the order of phase coherent length of electrons. Siegel has suggested on elegant method of categorizing nanophased materials in terms of modulation dimensionality (μ). In general, nanostructured materials are modulated over length scales of a few nm in zero, one, two or three dimensions – in addition to usually being spatially confined below 100 nm in one or more of these dimensions. Isolated clusters can be thought to have $\mu = 0$ (unmodulated), multilayers have $\mu = 1$ (i.e. modulated only in the direction normal to the layers), ultra-fine grained over layers have $\mu = 2$, while compacted nanomaterials have $\mu = 3$.

Synthesis

It is a great challenge to synthesize particles of nanometer dimensions with narrow size distribution and purity. A variety of methods are used to obtained nanostructured material. In principle, any method capable of producing very fine grain sized polycrystalline material can be utilized to produce nanocrystalline materials. Currently there are two approaches: reducing the size, also known as top down approach and building up by self-assembly or bottom up approach to obtain nano–size materials. In top down process larger materials are reduced to smaller dimensions of the order of nanometers and in bottom-up approach the materials are build-up to nanometer size. The commonly used methods can be categorized as physical and chemical methods.

Physical Method:

- Consolidation.
- Gas aggregation of monomer.
- Inert gas evaporation.
- Sputtering.
- Ion beam method.
- Ball milling.
- Lithography.

Chemical Methods:

- Chemical precipitation and capping.
- Sol-gel method.
- Micro-emulsion.
- Condensed phase synthesis.

- Reduction technique.
- Electro-chemical deposition.

Whatever be the method of synthesizing the nanoparticles, it is essential to avoid the coalescence of particles into larger particles and achieve monodispersity and chemical stability over a long period. Therefore the synthesis routes which protect or passivate the particles have become attractive compared to those which do not take care of long term stability or monodispersity.

All the above mentioned methods are used with different levels of success to produce nanocrystalline phases in a variety of materials. Usually gas condensation, mechanical alloying and chemical precipitation technique have been most commonly employed to produce nanocrystallites; while vapour deposition, sputtering and electrodeposition techniques have been used to synthesize the layered nanocrystals. The sol-gel process is generally used to produce clusters. Gas condensation and mechanical alloying are most commonly employed to produce large quantities of nanocrystallites.

Characterization

In order to understand the inter-relationship between structure and properties, nanocrystalline materials need to be characterized on both atomic and nanometer scales. The characteristic of above involves determining the shapes and sizes of nanoparticles and understanding of the inter-particle interactions. This information is important both from the scientific and the industrial application point of view. A number of experimental techniques have been employed to yield structural information on nanocrystalline materials. These include 'direct' microscopic techniques such as:

- Atomic force microscopy (AFM).
- Transmission electron microscopy (TEM).
- Scanning electron microscopy (SEM).
- Field ion microscopy (FIM).

In addition to these, many indirect techniques are also used to obtain the information about the nanomaterials indirectly. A few indirect techniques are:

- Absorption spectra.
- Diffraction of X-rays, electrons or neutrons.
- Rutherford back scattering (RBS).
- Raman spectroscopy.
- Auger electron spectroscopy (AES).
- Photoluminescence and Photoluminescence excitation.

X-ray diffraction patterns for various sizes of nanocrystals show that the peaks are broadened as the crystal size is reduced. The absorption edge or peak is shifted towards higher energy side by reducing the size indicating the widening of the forbidden energy gap of the material. These techniques are complementary to each other. Depending on the system to be studied, one technique may be better than the other.

Special Nanomaterials

With the advancement in technology, it has become possible to synthesize and characterize certain new types of materials having very small dimensions and novel characteristics. Nanomaterials are made by arranging the atom regularly at nanometer scale. This has given rise to certain special materials like fullerenes, carbon nanotubes, nanoporous materials, dendrimers, aerogels, zeolites, core–shell structure etc.

- Fullerenes: These are very large molecules of carbon having, 60, 70 or more atoms on the surface of nearly spherical cage like structure. These forms face centered cubic structure. It is possible to trap some ions inside the fullerene cage or attach some functional molecules to fullerene from outside. Due to their small size, these can be used for drug delivery.
- Carbon Nanotubes (CNTs): These can be considered as cylinders made of graphite sheets, mostly closed at ends. The thickness of sheet is just the atomic size of carbon atom and area about a few square micrometers. It is also possible that many concentric cylinders may form as a nanotube. Such nanotubes are termed as multiwall carbon nanotube (MWCNT). The distance between their walls is 0.334nm. These are more common; however under certain conditions it is possible to obtain single wall carbon nanotube (SWCNT). CNT have a great technological application in electronics, optoelectonics, drug delivery etc.
- Porous Silicon: Materials become porous when some solid particles are removed from them leaving voids in them. The properties of porous materials are quite different. Silicon is very widely used in electronics, but not much useful for optoelectronics due to its indirect band gap. Porous silicon with nano-size pores becomes direct band material and light emission is obtained from this in visible range. The nanosized pores in silicon wafers can be produced by chemical eteching, ion irradiation etc. This is useful for optical and electrical devices. Emission wavelength can be tuned by pore size.
- Aerogels: Aerogels constitute another class of highly porous materials. These ultra-low density materials are formed by inter connection of particles of nanometer size to form nano porous material. Pores themselves are usually non-uniform with sizes from ~ 10 to 100 nm. These materials are synthesized by solgel method and dried by specialized procedure to retain their porous structures. Aerogel formation of various materials such as silica, alumina, gelatin, cellulose and many others substances have been successfully demonstrated. These possess many

interesting properties because of their highly porous structure. They are lightest materials synthesized having pore diameter of few tens of nm with 2-5 nm size particles (80-90 % porous). These have low density (.003 to.08 g/cm^3), high surface area, low refractive index, low speed of sound through them, (~20 m/s), low Young's modulus and very low thermal conductivity (.003 W/m.k). Due to their special properties these find application as thermal insulators in jackets, blankets etc., sound proof rooms, filters for gaseous pollutants etc.

- Zeolites: Porous silicon and aerogels have disordered pores of irregular shapes and sizes. Zeolites are highly ordered porous materials. These have small pores (>2 nm) of uniform size which are periodically arranged to have long range order. The pore sizes can be controlled. The material is crystalline and useful to carry out different reactions. Tetrahederal units of Si and Al connected with Si- O- Al bonds is example of zeolites. These are useful catalysts or absorption materials. It is possible to synthesize and organize nanoparticles inside zeolites.
- Core shell particles: Core- shell particles form a novel class of nanocomposite materials in which a thin layer of nanometer size is coated on another material by some special procedure. The core can be just a nanoparticle (a few nm) with a nanometer thick coating or it can be a large core (few 100 nm) with nanometer thick coating. The properties are different for core or shell materials. These usually depend on core to shell ratio. Core shell particles can be synthesized using variety of combination of dielectrics, metal, semiconductor, dye, biomolecules etc. Core particles of different morphologies such as rods, rings, cube etc. also can be coated with thin shell. These can be utilized for creation of quantum bubbles (hollow sphere with thin shells).Core- shell structures are used for chemical stability, enhanced luminescence properties, engineering band structure, sensors, drug delivery etc.

Applications

Because of the very fine grain sizes, nanocrystalline materials exhibit a variety of properties that are different and often considerably improved in comparison with those of conventional coarse-grained polycrystalline materials. The fascinating field of nanotechnology has wide range of applications. Use of nanoscale materials may improve the performance of presently available devices. The nano-version of silica, titanium dioxide, clays, powdered metals, polymers and chemical products will be established in near future. New materials with difference performance characteristics may also be developed from nanostructured materials. On the other hand some new products have come up such as nanotubes, bucky balls, dendrimers, quantum dots etc. which are at research level.

A few of the applications are mentioned below:

- Tougher and harder cutting tools: Cutting tools made of nanocrystalline materials, such as tungsten carbide, tantalum carbide, and titanium carbide, are much harder, much more wear-resistant, erosion-resistant, and last longer than their conventional (large-grained) counterparts. They also enable the manufacturer to machine various materials much faster, thereby increasing productivity and significantly reducing manufacturing costs. Also, for the miniaturization of microelectronic circuits, the industry needs microdrills with enhanced edge retention and far better wear resistance. Since nanocrystalline carbides are much stronger, harder, and wear-resistant, they are currently being used in these microdrills.

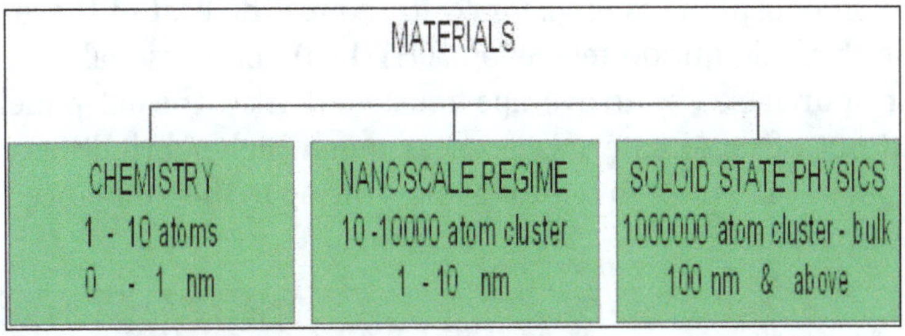

Size relationship of chemistry, nanoparticles and solid state physics.

- Ductile, machinable ceramics: Ceramics are very hard, brittle, and difficult to machine. These characteristics of ceramics have discouraged the potential users from exploiting their beneficial properties. However, with a reduction in grain size, these ceramics have increasingly been used. Zirconia, a hard, brittle ceramic, has even been rendered superplastic, i. e., it can be deformed to great lengths (up to 300% of its original length). However, these ceramics must possess nanocrystalline grains to be superplastic. In fact, nanocrystalline ceramics, such as silicon nitride (Si_3N_4) and silicon carbide (SiC), have been used in such automotive applications as high-strength springs, ball bearings, and valve lifters, because they possess good formability and machinabilty combined with excellent physical, chemical, and mechanical properties. They are also used as components in high-temperature furnaces. Nanocrystalline ceramics can be pressed and sintered into various shapes at significantly lower temperatures, whereas it would be very difficult, if not impossible, to press and sinter conventional ceramics even at high temperatures.

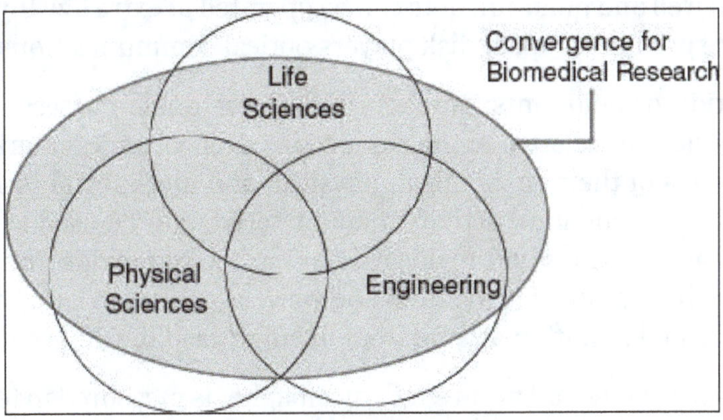

Intersection of scientific disciplines.

- Phosphors for high-definition TV & flat-panel displays: The resolution of a television, or a monitor, depends greatly on the size of the pixel. These pixels are essentially made of materials called "phosphors," which glow when struck by a stream of electrons inside the cathode ray tube (CRT). The resolution improves with a reduction in the size of the pixel, or the phosphors. Nanocrystalline zinc selenide, zinc sulfide, cadmium sulfide, and lead telluride synthesized by the sol-gel technique are candidates for improving the resolution of monitors. The use of nanophosphors is envisioned to reduce the cost of these displays so as to render high-definition televisions (HDTVs) and personal computers affordable to be purchased by an average household. Also, the flat-panel displays constructed out of nanomaterials possess much higher brightness, contrast and fast response than the conventional ones owing to their enhanced electrical and magnetic properties.

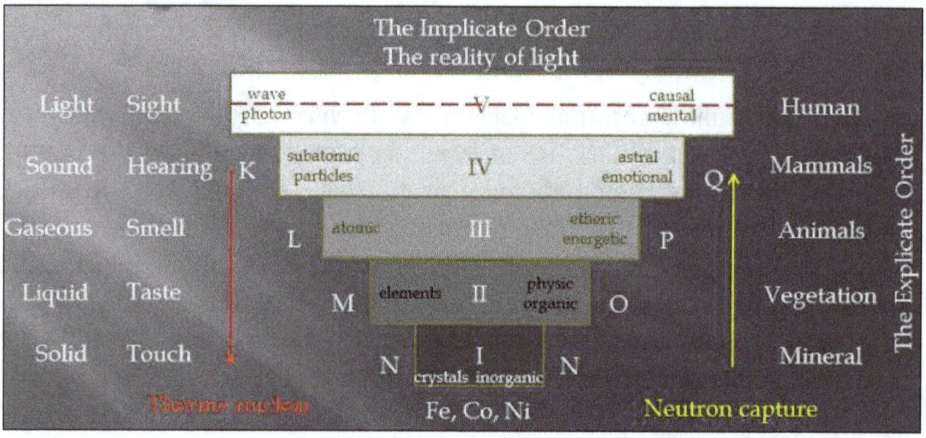

Scaling down Materials: Degree of confinement.

- Optical filters and attenuators: Optical transparency of nanocrystalline ceramics can be controlled by grain size. By changing the size optical filters for different colors can be made. These are also useful for sunscreens etc.
- Lasers with low threshold current: Double heterostructure lasers made from quantum well and quantum dots can be operated at extra low threshold current, which are used in compact disk players optical communications etc.
- Elimination of pollutants: Nanocrystalline materials possess extremely large grain boundaries relative to their grain size. Hence, nanomaterials are very active in terms of their of chemical, physical, and mechanical properties. Due to their enhanced chemical activity, nanomaterials can be used as catalysts to react with such noxious and toxic gases as carbon monoxide and nitrogen oxide in automobile catalytic converters and power generation equipment to prevent environmental pollution arising from burning gasoline and coal.
- High-power magnets: The strength of a magnet is measured in terms of coercivity and saturation magnetization values. These values increase with a decrease

in the grain size and an increase in the specific surface area (surface area per unit volume of the grains) of the grains. It has been shown that magnets made of nanocrystalline yttrium-samarium-cobalt grains possess very unusual magnetic properties due to their extremely large surface area. Typical applications for these high-power rare-earth magnets include quieter submarines, automobile alternators, land-based power generators, motors for ships, ultra-sensitive analytical instruments, and magnetic resonance imaging (MRI) in medical diagnostics.

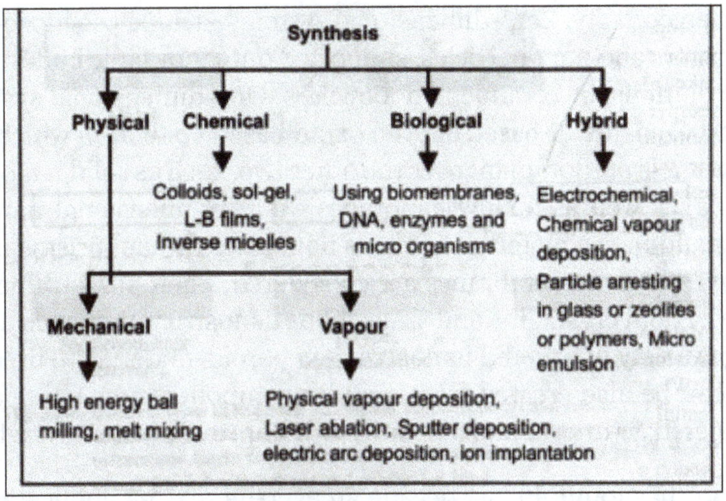

Methods of synthesis.

The emerging fields of nanoscience and nanoengineering are leading to unprecedented understanding and control over the fundamental building blocks of all physical things. Importance of nanotechnology is growing day by day because of its advantages in almost all fields like electronics, computers, medicine, engineering, defense etc. in addition to pure sciences. Nanoparticles nanotubes, nanofibres are changing today's technology because of their size dependent properties. Nanoporous materials are also playing an important role in nanotechnology because of some unusual characteristics. Many more applications may become possible in future due to novel and peculiar properties of nanostrcurues.

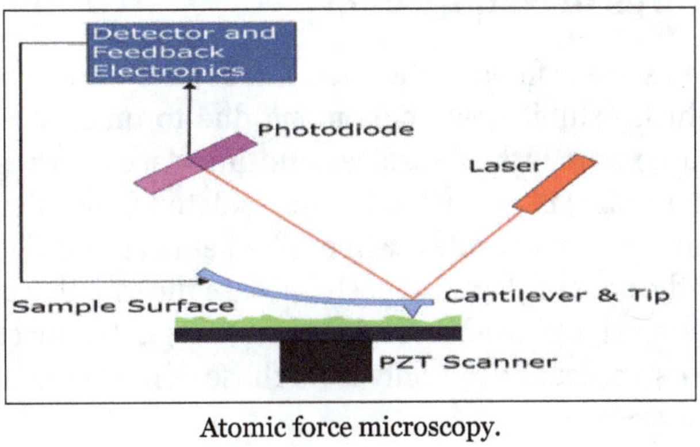

Atomic force microscopy.

NANOSYSTEMS

Inorganic nanosystems are defined as nanosized chemical objects whose composition is merely inorganic and which exhibit peculiar features due to quantum-size and geometrical effects. Two are the general synthetic pathways by which nano-objects are obtained: the so-called 'top-down' and 'bottom-up' approaches. Mainly in the latter method, chemistry plays a unique role in assembling and building up nanometric units from smaller ones. The nanosystems can be defined and classified according to the hierarchical order of dimensionality. Zero-dimensional systems include pseudospherical objects such as nanoclusters and nanoparticles, supported onto inorganic bulk supports as well as in colloidal solutions, or ceramic nanopowders. One-dimensional systems take into account carbon-based, metal-based or even oxide-based systems in which the extension over one dimension is predominant over the other two, such as solid nanofibers, nanowires or nanorods, as well as hollow nanotubes. As twodimensional nanosystems, the crystalline flat nanometric materials, such as nanodiscs or nanoprisms, and the amorphous nanofilms and nanomembranes are considered. Then, three-dimensional nanosystems consist of both crystalline and amorphous nanostructures, such as nanocrystals and a very large variety of ordered nanoarranged porous materials. Three-dimensional arrangements can be also created from simpler components, as nanoparticles or nanorods, and superstructures or superlattices with improved features are thus obtained.

The prefix "nano" has found in last decade an ever-increasing application to different fields of the knowledge. Nanoscience, nanotechnology, nanomaterials or nanochemistry are only a few of the new nano-containing terms that occur frequently in scientific reports, in popular books as well as in newspapers and that have become familiar to a wide public even non-experts.

Within the convention of International System of Units (SI) it is used to indicate a reduction factor of 109 times. So, the nanosized world is typically measured in nanometers (1 nm corresponding to 10-9 m) and it encompasses systems whose size is above molecular dimensions and below macroscopic ones (generally > 1 nm and < 100 nm).

Distinctive Features of Nanosystems

Inorganic nanosystems are defined as the chemical objects whose composition is merely inorganic and which exhibit new phenomena due to quantum-size effects and to the occurrence of large amounts of surfaces and interfaces because of their reduced size in the nanometer scale (1 - 100 nm). In fact, isolated molecules exhibit properties that follow quantum mechanical rules, while the chemical and physical properties of bulk materials obey the laws of classical mechanics. In the middle, nanosystems display electronic, photochemical, electrochemical, optical, magnetic, mechanical or catalytic properties that differ significantly not only from those of molecular units, but also from those of macroscopic systems.

Quantum-size effects arise in nanosized objects because their global dimensions are comparable to the characteristic wavelength for fundamental excitations in materials. These excitations (including the wavelength of electrons, photons and so on) carry the quanta of energy through materials and therefore govern the dynamics of their propagation and conversion from one form to another. However, if the size of the structures falls in the same order of magnitude of these characteristic wave functions, the propagation and the behavior of quanta are noticeably perturbed and thus quantum mechanical selection rules, which are not usually evident at larger scale, appear. For example, in the case of metals, typical "metallic" properties, like conductivity, decrease when the size is reduced and when the number of constituent atoms in the sample is significantly diminished. Indeed, the electronic conduction band of a metal gradually evolves from continuous levels of a bulk infinite material into discrete states as a function of size reduction, resulting in an increase in the band-gap energy.

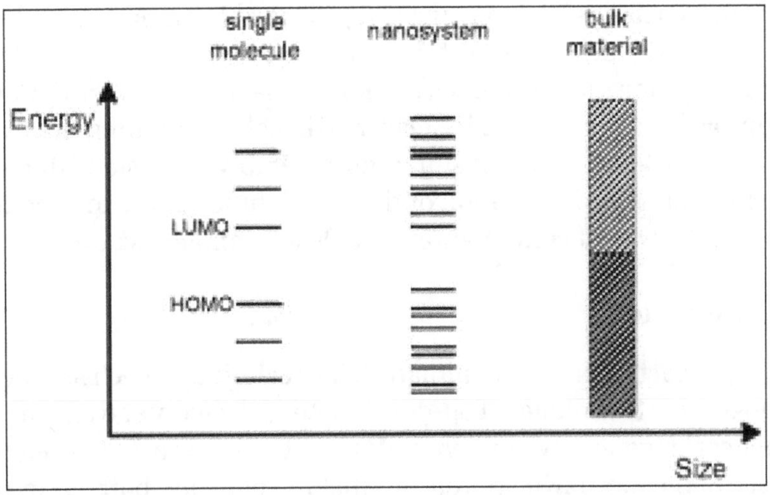

Nanoscale materials also exhibit size-dependent magnetic behavior. For instance, at very small sizes, magnetic nanoclusters have a single magnetic domain and the strongly coupled magnetic spins of each atom combine cooperatively to give rise to a system with a single "giant" spin.

In addition, nanosized materials possess very high surface to volume ratios because of the fine grain size and they are characterized by a very large amount of lowcoordination number atoms at edge and corner sites. Such preponderance of surfaces is a major reason for the change in chemical behavior of materials at the nanoscale. The atoms in nanostructures have a higher average energy than atoms in larger structures, because most of them are surface atoms. Consequently, the activity of a catalytic material can be exponentially improved as the catalyst is reduced in size at the nanoscale. In summary, in the nanometric world, the properties of nanosystems are remarkably affected by sometimes minor changes in size, shape or surface state of the structures. Indeed, three factors, namely (1) synthesis, (2) composition and structure and (3) functional properties, represent the essential relations in nanoscience.

General Aspects of Synthesis

Atoms and molecules are the essential building blocks of every object. The manner in which things are constructed with these basic units is vitally important to understand their properties and their reciprocal interactions. An efficient control of the synthetic pathways is essential during the preparation of nanobuilding blocks with different sizes and shapes that can lead to the creation of new devices and technologies with improved performances. To do this, two opposite, but complementary approaches are pursued. One is a top-down strategy of miniaturizing current components and materials, while the other is a bottom-up strategy of building ever-more-complex molecular structures atom by atom or molecule by molecule. These two different methods highlight the organization level of nanosystems as the crossing point hanging between the worlds of molecular objects and bulk materials.

The top-down approach has been advanced by Richard Feynman in his lecture stating that "there is plenty of room at the bottom" and it is ideal for obtaining structures with long-range order and for making connections with macroscopic world. Conversely, the bottom-up approach was pioneered by Jean-Marie Lehn (revealing that "there is plenty of room at the top") and it is best suited for assembly and establishing short-range order at the nanoscale. The integration of the two techniques is expected to provide, at least in principle, the widest combination of tools for nanofabrication.

Top-down Approach

The top-down approach is based on miniaturizing techniques, such as machining, templating or lithographic techniques. Top-down methods usually start from patterns generated at larger scale (generally at microscale) and then they are reduced to nanoscale. A key advantage of the top-down approach is that the parts are both patterned and built in place, so that no further assembly steps are needed.

By means of electronic, ionic or X-ray lithography, a monolith can be cut step by step in order to generate a quantum well (a bidimensional structure, with two finite dimensions) at first, then a quantum wire (a monodimensional structure, one finite dimension), and finally a quantum dot (a zero-dimensional structure, all the dimensions being in nanoscale). Current short-wavelength optical lithography methods can reach dimensions not less than 100 nanometers (the traditional threshold definition of the nanoscale). Extreme ultraviolet and X-ray sources are being developed to allow lithographic printing techniques to reach dimensions from 10 to 100 nanometers, but the principal limits are still due to the difficulty of beam focalization. Likewise, scanning beam techniques such as electron-beam lithography provide patterns down to about 20 nanometers and still-smaller features are obtained by using scanning probes to deposit or remove thin layers.

The general procedure of mechanical printing techniques consist on making a master "stamp" by a high-resolution lithographic technique, as described above, and then applying this stamp, or subsequent copies of it, to a surface to mould the pattern. The last step is to remove the thin layer of the masking material under the stamped regions. These nanoscale printing techniques offer several advantages due to the possibility to use a wide variety of materials with curved surfaces.

As a general drawback, these techniques are not cheap and require a complex manufacturing. In addition, top-down methodologies: (1), even if they work well at the microscale, they collide with some difficulties at nanoscale dimensions and (2) they usually lead to the formation of bidimensional structures and hence, as they are carried out by the addition or subtraction of patterned layers, they cannot easily give rise to the production of arbitrary three-dimensional objects.

It is worth underlining that the development in the top-down methodology was mainly driven from the traditional disciplines of materials engineering and physics, whereas the role of inorganic chemists has been minor in the exploitation of these techniques.

Bottom-up Approach

Bottom-up, or self-assembly, approaches to nanofabrication involve gradual additions of atoms or groups of atoms. This technique uses chemical or physical forces operating at the nanoscale to assemble basic units into larger structures. The chemical growth of nanometer-sized materials often implies colloidal or supramolecular systems and it frequently passes through phase transformations, such as vapor deposition on surfaces or precipitation of a solid phase from solution. Inspiration for bottom-up approaches comes from biological systems, where nature has employed chemical forces to create essentially all the structures needed by life. Researchers try to mimic nature's ability to produce small clusters of specific atoms, which can then self-assemble into more elaborated structures.

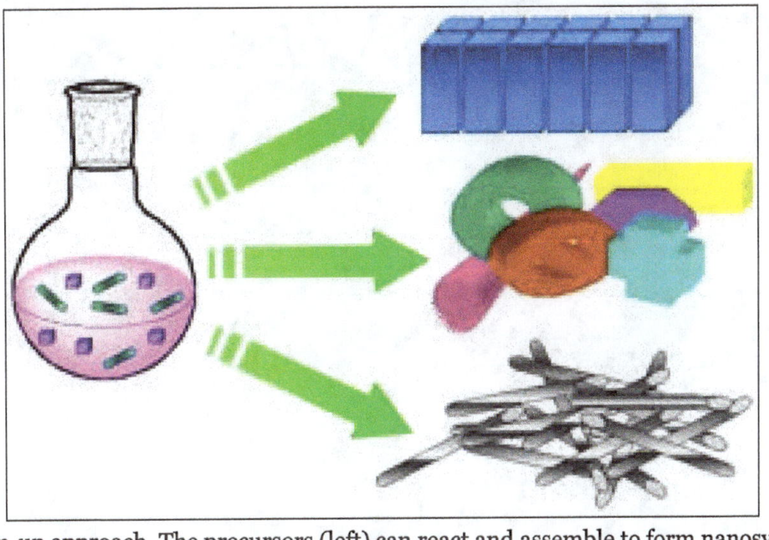

Bottom-up approach. The precursors (left) can react and assemble to form nanosystems with a large variety of shapes and sizes (right), depending on the reaction conditions.

In order to reach the desired shape and dimension of the new nanosystem, the nucleation and growth of the material have to be directed and controlled. This synthesis approach is of sure the most stimulating for the chemist. Indeed, bottom-up approaches, starting from single atoms and molecules have more affinity with chemistry and molecular biology, as much of chemistry already implicates the control of nanodimensional objects or the self-assembly of molecules into larger structures.

As main advantages, bottom-up techniques display a wide variety of preparation methods, they allow a good control onto scale dimension, even from atomic or molecular level, and they are not as expensive as top-down approaches. Nevertheless, the advent of pre-programmed self-assembling of arbitrarily large systems, with complexity comparable to that found in natural systems, is still a challenge.

References

- Nanoengineering: nanowerk.com, Retrieved 02 June, 2020
- Nanotechnology, technology: britannica.com, Retrieved 28 August, 2020
- Nanochemistry A Chemical Approach to Nanomaterials, RSC Publishing
- Light Scattering by Nonspherical Particles: Theory, Measurements and Applications, ed. M. I. Mishchenko, J. W. Hovenier and L. D. Travis, Academic Press, San Diego

2

Branches of Nanotechnology

Nanotechnology is categorized into a varied number of branches. These include molecular nanotechnology, nanomechanics, nanoarchitectonics, nanoinformatics, nanometrology, nanofluidics, etc. This chapter has been carefully written to provide an easy understanding of the different branches of nanotechnology.

MOLECULAR NANOTECHNOLOGY

Molecular Nanotechnology (MNT) is a technology that is used to design complex structures through mechanosynthesis process, in order to obtain the correct atomic specifications. In this technology, complex products are built using nanomachines. This process is not at all similar to nanomaterials as it is based on molecular manufacturing. The mechanosynthesis process used for this technology will be assisted by complex molecular machine systems.

The whole process of molecular nanotechnology would include the combination of physical theories with chemical demonstrations and other nanotechnologies, in addition to the different methods used in macro scale factories.

Unlike any other conventional processes, in MNT a clear and definite process is carried out to get a definite result.

MNT is actually a multi-product reaction process in which the molecular reactions are to be perfectly balanced in order to obtain the considered chemical reaction. The products of these reactions are again used to build a whole new system.

Applications of Molecular Nanotechnology

- Smart Materials: Smart materials are products that are made at nanometer scale for doing the desired task. It has already found great applications in the field of nanotechnology. It is useful in providing the correct response when it comes in contact with different molecules.

  Thus, artificial drugs could be produced which would easily find the specific viruses and delete them. Another idea of smart materials is self-healing structures which can be used for removing small tears from the surface just like the human skin.

- Nanosensors: A nanosensor can also be called as a smart material. The only small difference is that a nano sized component will be governed by a larger machine. This machine responds to its surroundings and brings the intended changes. The best example of a nanosensor is the photosensor. The device is used to measure the light that hits on its surface and convert it into its corresponding electricity and is passes onto another bigger device. Such a sensor when converted to nanosize, will cost much lesser than the conventional ones, and will also use lesser power. But, their applications become limited.
- Replicating Nanorobots: The main idea behind molecular nanotechnology is to create hundreds of nanorobots which work together in coordination. These nanorobots should also be able to design and produce more nanorobots with the help of sophisticated building blocks in an artificial environment.

  There has always been a doubt in mind about how feasible the concept of nanorobots and its replicating features would be. There is a fear of difference in characteristics with the parent robots through replication.

  Nanorobots have also pioneered in the medical field. When a number of nanorobots are designed to work in the medical field, a number of patients can be taken care of simultaneously and that too for hours. These robots can also help in correcting certain genetic defects and will also help in increasing the life span of people. These nanorobots are may be made to mirror human movements, but are not allowed to self-replicate inside the human body.
- Utility fog: This is also one basic application of MNT in which a series of nanorobots are networked so that they can change their characteristics according to the change in the commands of the software. Through this method, the physical objects that are used for multitasks can be replaced by nanorobots.

Advantages of MNT

In future, MNT will help in making highly accurate devices like computers and that too in nanosizes.

The products made by MNT's will be stronger and more powerful than the current devices that are made out of iron or other alloys. All the necessary objects in our everyday life like chairs, cars and so on can be made in cheap rates with multiple times efficiency, lesser weight and more durability.

As these devices are minute is size, they can be used to make small surgical tools that can be injected into our body in order to search and destroy cancer cells and other virus infections. They can also be used to remove clots and also provide oxygen when circulation problems occur.

It is sure that nanotechnology will bring about a whole new perspective of making products. The devices that we use every day will become more accurate and every product in this world will have a new look.

Concerns Regarding Molecular Nanotechnology

- One of the main risks concerned with MNT is technological singularity. That is, the good things about the technology could be used for bad stuffs like creating deadly weapons. These weapons could easily self-replicate and cause chaos in the world. Thus, protocols must be taken as to the permit of such illegal weapons and their self-replication.
- One another concern is that if the nanorobots can self-replicate to a high level, the chances of them consuming the whole world for raw materials is high. Such a problem is generally termed as grey goo.

NANOBIOTECHNOLOGY

Nanobiotechnology, bionanotechnology, and nanobiology are terms that refer to the intersection of nanotechnology and biology. Given that the subject is one that has only emerged very recently, bionanotechnology and nanobiotechnology serve as blanket terms for various related technologies.

This discipline helps to indicate the merger of biological research with various fields of nanotechnology. Concepts that are enhanced through nanobiology include: nanodevices (such as biological machines), nanoparticles, and nanoscale phenomena that occurs within the discipline of nanotechnology. This technical approach to biology allows scientists to imagine and create systems that can be used for biological research. Biologically inspired nanotechnology uses biological systems as the inspirations for technologies not yet created. However, as with nanotechnology and biotechnology, bionanotechnology does have many potential ethical issues associated with it.

The most important objectives that are frequently found in nanobiology involve applying nanotools to relevant medical/biological problems and refining these applications. Developing new tools, such as peptoid nanosheets, for medical and biological purposes is another primary objective in nanotechnology. New nanotools are often made by refining the applications of the nanotools that are already being used. The imaging of native biomolecules, biological membranes, and tissues is also a major topic for nanobiology researchers. Other topics concerning nanobiology include the use of cantilever array sensors and the application of nanophotonics for manipulating molecular processes in living cells.

Recently, the use of microorganisms to synthesize functional nanoparticles has been of great interest. Microorganisms can change the oxidation state of metals. These microbial processes have opened up new opportunities for us to explore novel applications, for example, the biosynthesis of metal nanomaterials. In contrast to chemical and physical methods, microbial processes for synthesizing nanomaterials can be achieved in aqueous phase under gentle and environmentally benign conditions. This approach has become an attractive focus in current green bionanotechnology research towards sustainable development.

The terms are often used interchangeably. When a distinction is intended, though, it is based on whether the focus is on applying biological ideas or on studying biology with nanotechnology. Bionanotechnology generally refers to the study of how the goals of nanotechnology can be guided by studying how biological "machines" work and adapting these biological motifs into improving existing nanotechnologies or creating new ones. Nanobiotechnology, on the other hand, refers to the ways that nanotechnology is used to create devices to study biological systems.

In other words, nanobiotechnology is essentially miniaturized biotechnology, whereas bionanotechnology is a specific application of nanotechnology. For example, DNA nanotechnology or cellular engineering would be classified as bionanotechnology because they involve working with biomolecules on the nanoscale. Conversely, many new medical technologies involving nanoparticles as delivery systems or as sensors would be examples of nanobiotechnology since they involve using nanotechnology to advance the goals of biology.

The definitions enumerated above will be utilized whenever a distinction between nanobio and bionano is made in this article. However, given the overlapping usage of the terms in modern parlance, individual technologies may need to be evaluated to determine which term is more fitting. As such, they are best discussed in parallel.

Most of the scientific concepts in bionanotechnology are derived from other fields. Biochemical principles that are used to understand the material properties of biological systems are central in bionanotechnology because those same principles are to be used to create new technologies. Material properties and applications studied in bionanoscience include mechanical properties (e.g. deformation, adhesion, failure), electrical/electronic (e.g. electromechanical stimulation, capacitors, energy storage/batteries), optical (e.g. absorption, luminescence, photochemistry), thermal (e.g. thermomutability, thermal management), biological (e.g. how cells interact with nanomaterials, molecular flaws/defects, biosensing, biological mechanisms such as mechanosensation), nanoscience of disease (e.g. genetic disease, cancer, organ/tissue failure), as well as computing (e.g. DNA computing) and agriculture (target delivery of pesticides, hormones and fertilizers. The impact of bionanoscience, achieved through structural and mechanistic analyses of biological processes at nanoscale, is their translation into synthetic and technological applications through nanotechnology.

Nanobiotechnology takes most of its fundamentals from nanotechnology. Most of the devices designed for nano-biotechnological use are directly based on other existing nanotechnologies. Nanobiotechnology is often used to describe the overlapping multidisciplinary activities associated with biosensors, particularly where photonics, chemistry, biology, biophysics, nanomedicine, and engineering converge. Measurement in biology using wave guide techniques, such as dual-polarization interferometry, is another example.

Applications

Applications of bionanotechnology are extremely widespread. Insofar as the distinction holds, nanobiotechnology is much more commonplace in that it simply provides more tools for the study of biology. Bionanotechnology, on the other hand, promises to recreate biological mechanisms and pathways in a form that is useful in other ways.

Nanomedicine

Nanomedicine is a field of medical science whose applications are increasing more and more thanks to nanorobots and biological machines, which constitute a very useful tool to develop this area of knowledge. In the past years, researchers have made many improvements in the different devices and systems required to develop nanorobots. This supposes a new way of treating and dealing with diseases such as cancer; thanks to nanorobots, side effects of chemotherapy have been controlled, reduced and even eliminated, so some years from now, cancer patients will be offered an alternative to treat this disease instead of chemotherapy, which causes secondary effects such as hair loss, fatigue or nausea killing not only cancerous cells but also the healthy ones. At a clinical level, cancer treatment with nanomedicine will consist of the supply of nanorobots to the patient through an injection that will search for cancerous cells while leaving the healthy ones untouched. Patients that will be treated through nanomedicine will not notice the presence of these nanomachines inside them; the only thing that is going to be noticeable is the progressive improvement of their health. Nanobiotechanlogy is quite important for medicine formulation. It helps a lot in making vaccines as well.

Nanobiotechnology

Nanobiotechnology (sometimes referred to as nanobiology) is best described as helping modern medicine progress from treating symptoms to generating cures and regenerating biological tissues. Three American patients have received whole cultured bladders with the help of doctors who use nanobiology techniques in their practice. Also, it has been demonstrated in animal studies that a uterus can be grown outside the body and then placed in the body in order to produce a baby. Stem cell treatments have been used to fix diseases that are found in the human heart and are in clinical trials in the United States. There is also funding for research into allowing people to have new limbs without having to resort to prosthesis. Artificial proteins might also become available to manufacture without the need for harsh chemicals and expensive machines. It has even been surmised that by the year 2055, computers may be made out of biochemicals and organic salts.

Another example of current nanobiotechnological research involves nanospheres coated with fluorescent polymers. Researchers are seeking to design polymers whose fluorescence is quenched when they encounter specific molecules. Different polymers would detect different metabolites. The polymer-coated spheres could become part of new biological assays, and the technology might someday lead to particles which could be introduced into the human body to track down metabolites associated with tumors and other health problems. Another example, from a different perspective, would be evaluation and therapy at the nanoscopic level, i.e. the treatment of Nanobacteria (25-200 nm sized) as is done by NanoBiotech Pharma.

While nanobiology is in its infancy, there are a lot of promising methods that will rely on nanobiology in the future. Biological systems are inherently nano in scale; nanoscience must merge with biology in order to deliver biomacromolecules and molecular machines that are similar to nature. Controlling and mimicking the devices and processes that are constructed from molecules is a tremendous challenge to face for the converging disciplines of nanobiotechnology. All living things, including humans, can be considered to be nanofoundries. Natural evolution has optimized the "natural" form of nanobiology over millions of years. In the 21st century, humans have developed the technology to artificially tap into nanobiology. This process is best described as "organic merging with synthetic." Colonies of live neurons can live together on a biochip device; according to research from Dr. Gunther Gross at the University of North Texas. Self-assembling nanotubes have the ability to be used as a structural system. They would be composed together with rhodopsins; which would facilitate the optical computing process and help with the storage of biological materials. DNA (as the software for all living things) can be used as a structural proteomic system - a logical component for molecular computing. Ned Seeman - a researcher at New York University - along with other researchers are currently researching concepts that are similar to each other.

Bionanotechnology

DNA nanotechnology is one important example of bionanotechnology. The utilization of the inherent properties of nucleic acids like DNA to create useful materials is a promising area of modern research. Another important area of research involves taking advantage of membrane properties to generate synthetic membranes. Proteins that self-assemble to generate functional materials could be used as a novel approach for the large-scale production of programmable nanomaterials. One example is the development of amyloids found in bacterial biofilms as engineered nanomaterials that can be programmed genetically to have different properties. Protein folding studies provide a third important avenue of research, but one that has been largely inhibited by our inability to predict protein folding with a sufficiently high degree of accuracy.

Given the myriad uses that biological systems have for proteins, though, research into understanding protein folding is of high importance and could prove fruitful for bionanotechnology in the future.

Lipid nanotechnology is another major area of research in bionanotechnology, where physico-chemical properties of lipids such as their antifouling and self-assembly is exploited to build nanodevices with applications in medicine and engineering.

Agriculture

Nanotechnology application to biotechnology leaves no field untouched by its groundbreaking scientific innovations for human wellness; the agricultural industry is no exception. Basically, nanomaterials are distinguished depending on the origin: natural, incidental and engineered nanoparticles. Among these, engineered nanoparticles have received wide attention in all fields of science, including medical, materials and agriculture technology with significant socio-economical growth. In the agriculture industry, engineered nanoparticles have been serving as nano carriers, containing herbicides, chemicals, or genes, which target particular plant parts to release their content. Previously nanocapsules containing herbicides have been reported to effectively penetrate through cuticles and tissues, allowing the slow and constant release of the active substances. Likewise, other literature describes that nano-encapsulated slow release of fertilizers has also become a trend to save fertilizer consumption and to minimize environmental pollution through precision farming.

These are only a few examples from numerous research works which might open up exciting opportunities for nanobiotechnology application in agriculture. Also, application of this kind of engineered nanoparticles to plants should be considered the level of amicability before it is employed in agriculture practices. Based on a thorough literature survey, it was understood that there is only limited authentic information available to explain the biological consequence of engineered nanoparticles on treated plants. Certain reports underline the phytotoxicity of various origin of engineered nanoparticles to the plant caused by the subject of concentrations and sizes. At the same time, however, an equal number of studies were reported with a positive outcome of nanoparticles, which facilitate growth promoting nature to treat plant. In particular, compared to other nanoparticles, silver and gold nanoparticles based applications elicited beneficial results on various plant species with less and/or no toxicity. Silver nanoparticles (AgNPs) treated leaves of Asparagus showed the increased content of ascorbate and chlorophyll. Similarly, AgNPs-treated common bean and corn has increased shoot and root length, leaf surface area, chlorophyll, carbohydrate and protein contents reported earlier. The gold nanoparticle has been used to induce growth and seed yield in Brassica juncea.

Tools

This field relies on a variety of research methods, including experimental tools (e.g. imaging, characterization via AFM/optical tweezers etc.), x-ray diffraction based tools, synthesis via self-assembly, characterization of self-assembly (using e.g. MP-SPR, DPI, recombinant DNA methods, etc.), theory (e.g. statistical mechanics, nanomechanics, etc.), as well as computational approaches (bottom-up multi-scale simulation, supercomputing).

NANOMECHANICS

Nanomechanics is a branch of *nanoscience* studying fundamental *mechanical* (elastic, thermal and kinetic) properties of physical systems at the nanometer scale. Nanomechanics has emerged on the crossroads of biophysics, classical mechanics, solid-state physics, statistical mechanics, materials science, and quantum chemistry. As an area of nanoscience, nanomechanics provides a scientific foundation of nanotechnology.

Nanomechanics is that branch of nanoscience which deals with the study and application of fundamental mechanical properties of physical systems at the nanoscale, such as elastic, thermal and kinetic material properties.

Often, nanomechanics is viewed as a branch of nanotechnology, i.e., an applied area with a focus on the mechanical properties of engineered nanostructures and nanosystems (systems with nanoscale components of importance). Examples of the latter include nanomachines, nanoparticles, nanopowders, nanowires, nanorods, nanoribbons, nanotubes, including carbon nanotubes (CNT) and boron nitride nanotubes (BNNTs); nanoshells, nanomebranes, nanocoatings, nanocomposite/nanostructured materials, (fluids with dispersed nanoparticles); nanomotors, etc.

Some of the well-established *fields of nanomechanics* are: nanomaterials, nanotribology (friction, wear and contact mechanics at the nanoscale), nanoelectromechanical systems (NEMS), and nanofluidics.

As a fundamental science, nanomechanics is based on some empirical principles (basic observations), namely general mechanics principles and specific principles arising from the smallness of physical sizes of the object of study.

General mechanics principles include:

- Energy and momentum conservation principles.
- Variational Hamilton's principle.
- Symmetry principles.

Due to smallness of the studied object, nanomechanics also accounts for:

- Discreteness of the object, whose size is comparable with the interatomic distances.

- Plurality, but finiteness, of degrees of freedom in the object.
- Importance of thermal fluctuations.
- Importance of entropic effects.
- Importance of quantum effects.

These principles serve to provide a basic insight into novel mechanical properties of nanometer objects. Novelty is understood in the sense that these properties are not present in similar macroscale objects or much different from the properties of those (e.g., nanorods vs. usual macroscopic beam structures). In particular, smallness of the subject itself gives rise to various surface effects determined by higher surface-to-volume ratio of nanostructures, and thus affects mechanoenergetic and thermal properties (melting point, heat capacitance, etc.) of nanostructures. Discreteness serves a fundamental reason, for instance, for the dispersion of mechanical waves in solids, and some special behavior of basic elastomechanics solutions at small scales. Plurality of degrees of freedom and the rise of thermal fluctuations are the reasons for thermal tunneling of nanoparticles through potential barriers, as well as for the cross-diffusion of liquids and solids. Smallness and thermal fluctuations provide the basic reasons of the Brownian motion of nanoparticles. Increased importance of thermal fluctuations and configuration entropy at the nanoscale give rise to superelasticity, entropic elasticity (entropic forces), and other exotic types of elasticity of nanostructures. Aspects of configuration entropy are also of great interest in the context self-organization and cooperative behavior of open nanosystems.

Quantum effects determine forces of interaction between individual atoms in physical objects, which are introduced in nanomechanics by means of some averaged mathematical models called interatomic potentials.

Subsequent utilization of the interatomic potentials within the classical multibody dynamics provide deterministic mechanical models of nano structures and systems at the atomic scale/resolution. Numerical methods of solution of these models are called molecular dynamics (MD), and sometimes molecular mechanics (especially, in relation to statically equilibrated (still) models). Non-deterministic numerical approaches include Monte Carlo, Kinetic More-Carlo (KMC), and other methods. Contemporary numerical tools include also hybrid multiscale approaches allowing concurrent or sequential utilization of the atomistic scale methods (usually, MD) with the continuum (macro) scale methods (usually, field emission microscopy) within a single mathematical model. Development of these complex methods is a separate subject of applied mechanics research.

Quantum effects also determine novel electrical, optical and chemical properties of nanostructures, and therefore they find even greater attention in adjacent areas of nanoscience and nanotechnology, such as nanoelectronics, advanced energy systems, and nanobiotechnology.

NANOBIOMECHANICS

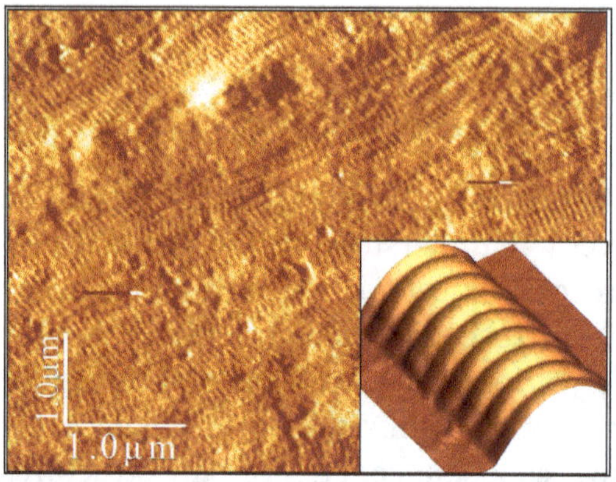

High resolution AFM image of cortical bone and single collagen fibril (inset).

Nanobiomechanics (also bionanomechanics) is an emerging field in nanoscience and biomechanics that combines the powerful tools of nanomechanics to explore fundamental science of biomaterials and biomechanics.

Since the introduction by its founder Yuan-Cheng Fung, the field of biomechanics has become one of the branches of mechanics and bioscience. For many years, biomechanics has examined tissue. Through advancements in nanoscience, the scale of the forces that could be measured and also the scale of observation of biomaterials was reduced to "nano" and "pico" level. Consequently, it became possible to measure the mechanical properties of biological materials at nanoscale.

Most of the biological materials have different hierarchical levels, and the smallest ones refer to the nanoscale. For example, bone has up to seven levels of biological organization, and the smallest level, i.e., single collagen fibril and hydroxylapatite minerals have dimensions well below 100 nm. Therefore, being able to probe properties at this small scales provides a great opportunity for better understanding the fundamental properties of these materials. For example, measurements have shown that nanomechanical heterogeneity exists even within single collagen fibrils as small as 100 nm.

One of the other most relevant topics in this field is measurement of tiny forces on living cells to recognize changes caused by different diseases. For example, it has been shown that red blood cells infected by malaria are 10 times stiffer than normal cells. Likewise, it has been shown that cancer cells are 70 percent softer than normal cells. Early signs of aging cartilage and osteoarthritis has been shown by looking at the changes in the tissue at the nanoscale.

Methods and Instrumentation

The common methods in nanobiomechanics are atomic force microscope, optical tweezers, and magnetic twisting cytometry.

Examples of relevant materials are bone and its hierarchical constituents such as single collagen fibrils, single living cells, actin filaments and microtubules, and synthetic peptide nanotubes.

Computational Nanobiomechanics

In addition to experimental aspect, research has been expanding through computational methods. Molecular dynamics (MD) simulations have provided a wealth of knowledge in this area. Although, the MD simulation are still limited to a small number of atoms and molecules, due to limitation in the computational performance, they have proved to be an instrumental branch of this emerging field.

NANOARCHITECTONICS

Nanoarchitectonics is a technology allowing to arrange nano-sized structural units, usually a group of atoms or molecules, in an intended configuration. It employs two major processes: Nano-creation and nano-organization. Nano-organization involves re-arrangement of the structural units in a desired pattern, while nano-creation is synthesis of new materials that do not exist in nature. For example, by peeling atomic sheets off graphite slab, a novel nano-material graphene can be obtained, which has very different properties from graphite.

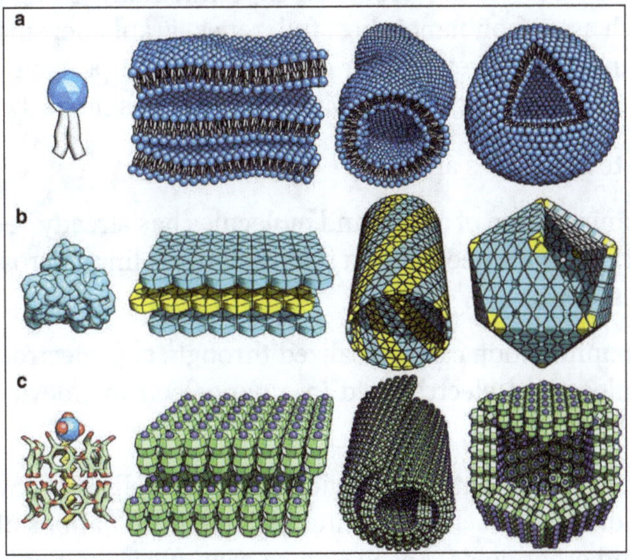

Nanoarchitectonics in biology: self-assembly of lipids (a), proteins (b), and (c) SDS-cyclodextrin complexes. SDS is a surfactant with a hydrocarbon tail (yellow) and a SO_4 head (blue and red), while cyclodextrin is a saccharide ring (green C and red O atoms).

Nanoarchitectonics is not limited to nano-creation and nano-organization, but rather employs those techniques to understand and use the ultimate functions of materials.

The important technologies to achieve this goal involve manipulation of single atoms and molecules through physical interactions, chemical reactions, applied fields, or self-assembly.

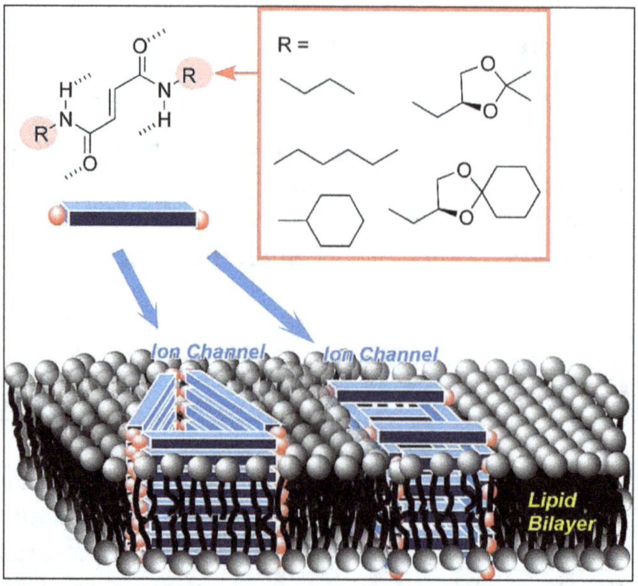

Nanoarchitectonics in chemistry: transmembrane chloride channel formed by self-assembly of fumaramides.

A typical example of nano-organization is the development of a nanoelectronics circuit. Challenging electronic devices are produced experimentally, using previously discovered materials, such as carbon nanotubes, fullerenes, graphene, single molecules having functional groups, etc. However, their practical use is impossible without a technology (nano-organization) to integrate and link these devices into a system.

Examples of those technologies are:

- Physical manipulation of atoms and molecules has already been achieved using, e.g., atomically sharp needles of a scanning tunneling microscope or an atomic force microscope.
- Chemical manipulation can be realized through, e.g., electrochemical reactions induced by localized electric field in a nanoelectronic device or through local polymerization.
- Field-induced manipulation is a widely explored direction where control over atoms or molecules is achieved using various combinations of electric, magnetic, elastic and other fields. A well-known example is manipulation of individual atoms by laser beams ("optical tweezers").
- Self-assembly usually involves weaker interactions, such as van der Waals forces. Common self-assembly examples are growth of a molecular crystal, e.g., snow.

The importance of architectonics in nanoscience and nanotechnology was first accentuated in 1999 at the University of California, Los Angeles (UCLA). Next year, the first International Symposium on Nanoarchitectonics has been held in Tsukuba, Japan. In 2003, the term nanoarchitectonics was first mentioned in a peer-reviewed article, by a German group from the Freie Universität Berlin, and the same year a dedicated research center, Functional Engineered Nano Architectonics, has been opened at UCLA. In 2007, similar centers have been established in Tsukuba: the International Center for Materials Nanoarchitectonics at NIMS and Interfacial Nanoarchitectonics at AIST.

NANOINFORMATICS

Nanoinformatics is the application of informatics to nanotechnology. It is an interdisciplinary field that develops methods and software tools for understanding nanomaterials, their properties, and their interactions with biological entities, and using that information more efficiently. It differs from cheminformatics in that nanomaterials usually involve nonuniform collections of particles that have distributions of physical properties that must be specified. The nanoinformatics infrastructure includes ontologies for nanomaterials, file formats, and data repositories.

Nanoinformatics has applications for improving workflows in fundamental research, manufacturing, and environmental health, allowing the use of high-throughput data-driven methods to analyze broad sets of experimental results. Nanomedicine applications include analysis of nanoparticle-based pharmaceuticals for structure–activity relationships in a similar manner to bioinformatics.

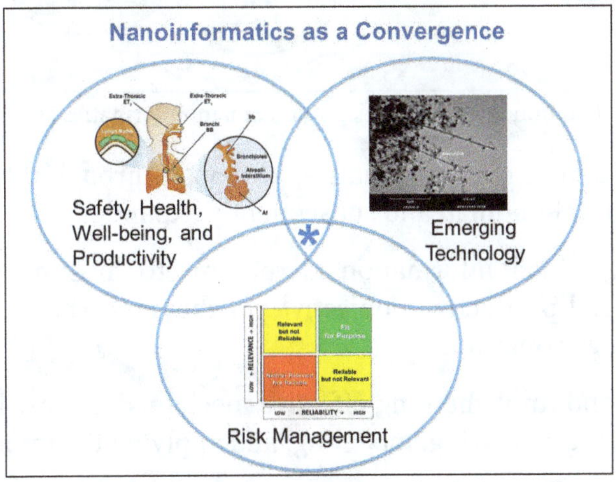

Context of nanoinformatics as a convergence of science and practice at the nexus of safety, health, well-being, and productivity; risk management; and emerging nanotechnology.

While conventional chemicals are specified by their chemical composition, and concentration, nanoparticles have other physical properties that must be measured for a complete description, such as size, shape, surface properties, crystallinity, and dispersion state. In addition, preparations of nanoparticles are often non-uniform, having distributions of these properties that must also be specified. These molecular-scale properties influence their macroscopic chemical and physical properties, as well as their biological effects. They are important in both the experimental characterization of nanoparticles and their representation in an informatics system. The context of nanoinformatics is that effective development and implementation of potential applications of nanotechnology requires the harnessing of information at the intersection of safety, health, well-being, and productivity; risk management; and emerging nanotechnology.

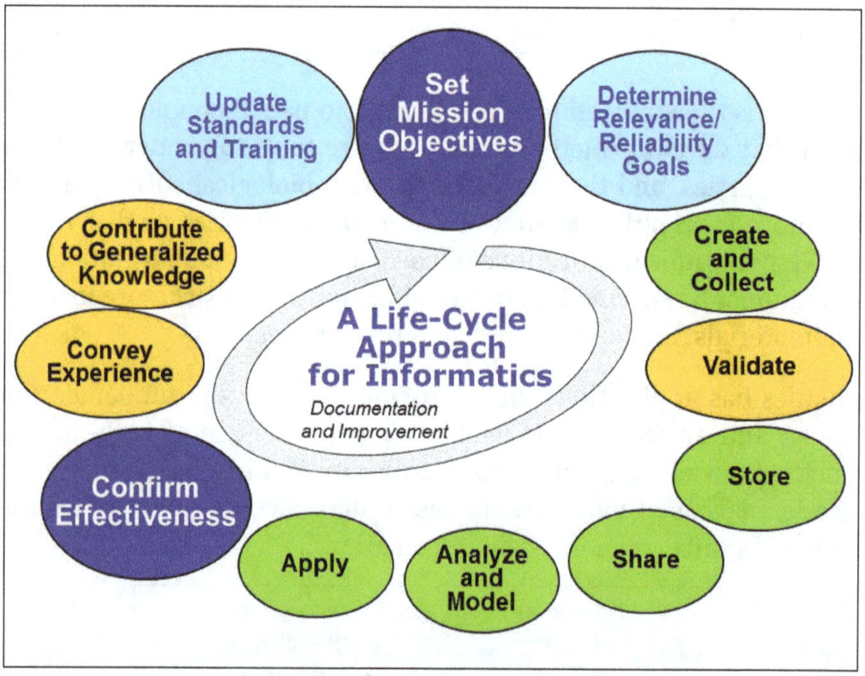

A graphical representation of a working definition of nanoinformatics as a life-cycle process.

One working definition of nanoinformatics developed through the community-based Nanoinformatics 2020 Roadmap and subsequently expanded is:

- Determining which information is relevant to meeting the safety, health, well-being, and productivity objectives of the nanoscale science, engineering, and technology community.
- Developing and implementing effective mechanisms for collecting, validating, storing, sharing, analyzing, modeling, and applying the information.
- Confirming that appropriate decisions were made and that desired mission outcomes were achieved as a result of that information.
- Conveying experience to the broader community, contributing to generalized knowledge, and updating standards and training.

Data Representations

Although nanotechnology is the subject of significant experimentation, much of the data are not stored in standardized formats or broadly accessible. Nanoinformatics initiatives seek to coordinate developments of data standards and informatics methods.

Ontologies

In the context of information science, an ontology is a formal representation of knowledge within a domain, using hierarchies of terms including their definitions, attributes, and relations. Ontologies provide a common terminology in a machine-readable framework that facilitates sharing and discovery of data. Having an established ontology for nanoparticles is important for cancer nanomedicine due to the need of researchers to search, access, and analyze large amounts of data.

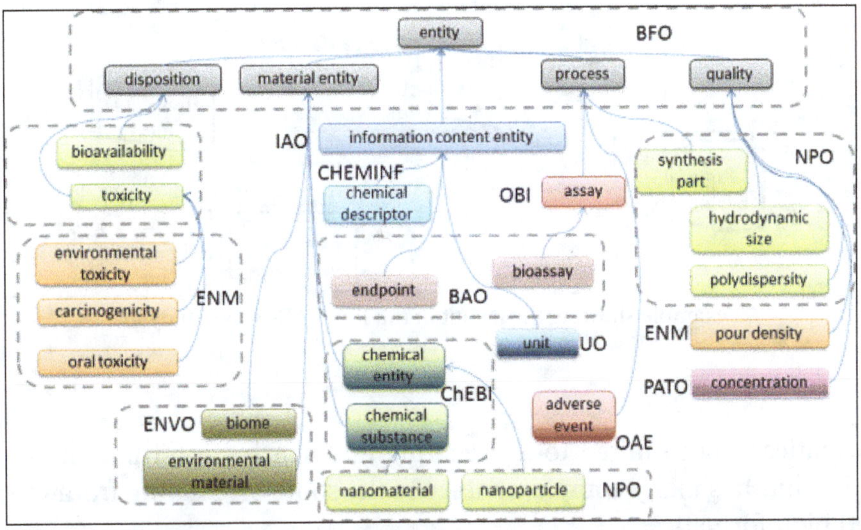

An overview of the eNanoMapper nanomaterial ontology.

The NanoParticle Ontology is an ontology for the preparation, chemical composition, and characterization of nanomaterials involved in cancer research. It uses the Basic Formal Ontology framework and is implemented in the Web Ontology Language. It is hosted by the National Center for Biomedical Ontology and maintained on GitHub. The eNanoMapper Ontology is more recent and reuses wherever possible already existing domain ontologies. As such, it reuses and extends the NanoParticle Ontology, but also the BioAssay Ontology, Experimental Factor Ontology, Unit Ontology, and ChEBI.

File Formats

ISA-TAB-Nano is a set of four spreadsheet-based file formats for representing and sharing nanomaterial data, based on the ISA-TAB metadata standard. In Europe, other templates have been adopted that were developed by the Institute of Occupational Medicine, and by the Joint Research Centre for the NANoREG project.

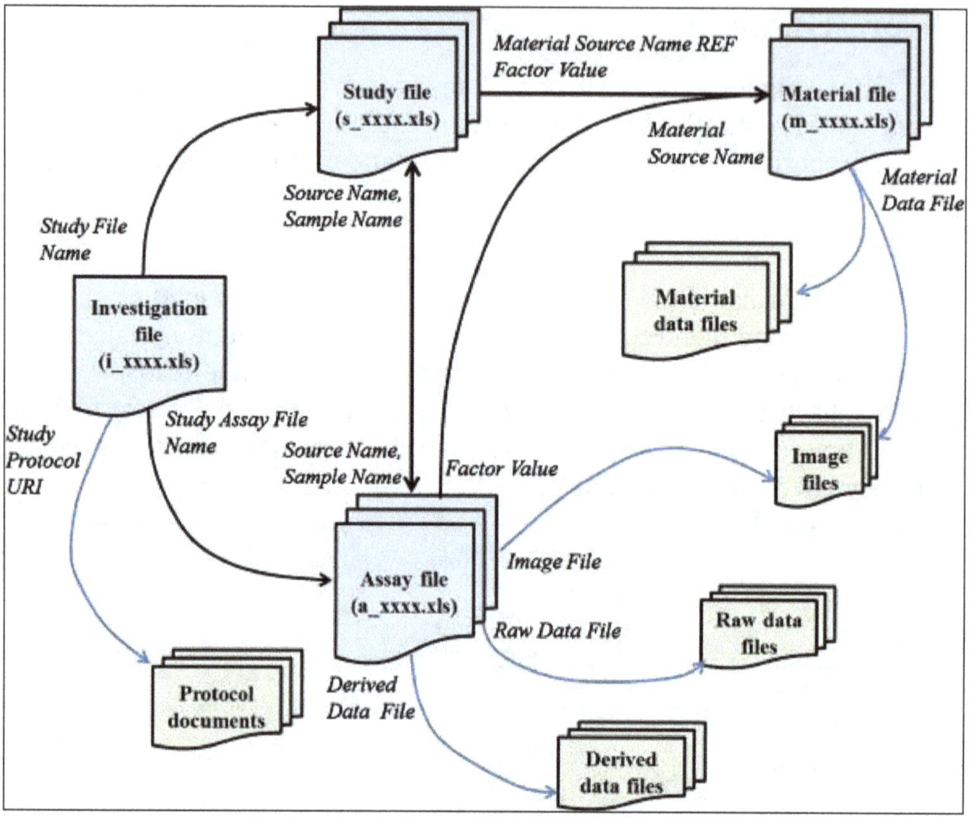

An example showing the layout of the ISA-TAB-Nano Study file.

Tools

Nanoinformatics is not limited to aggregating and sharing information about nanotechnologies, but has many complementary tools, some originating from chemoinformatics and bioinformatics.

Databases and Repositories

Over the last couple of years, various databases have been made available. caNanoLab, developed by the U.S. National Cancer Institute, focuses on nanotechnologies related to biomedicine. The NanoMaterials Registry, maintained by RTI International, is a curated database of nanomaterials, and includes data from caNanoLab.

The eNanoMapper database, a project of the EU NanoSafety Cluster, is a deployment of the database software developed in the eNanoMapper project. It has since been used in other settings, such as the EU Observatory for NanoMaterials (EUON). Other databases include the Center for the Environmental Implications of NanoTechnology's NanoInformatics Knowledge Commons (NIKC) and NanoDatabank, PEROSH's Nano Exposure & Contextual Information Database (NECID), Data and Knowledge on Nanomaterials (DaNa), and Springer Nature's Nano database.

Applications

Nanoinformatics has applications for improving workflows in fundamental research, manufacturing, and environmental health, allowing the use of high-throughput data-driven methods to analyze broad sets of experimental results.

Nanoinformatics is especially useful in nanoparticle-based cancer diagnostics and therapeutics. They are very diverse in nature due to the combinatorially large numbers of chemical and physical modifications that can be made to them, which can cause drastic changes in their functional properties. This leads to a combinatorial complexity that far exceeds, for example, genomic data. Nanoinformatics can enable structure–activity relationship modelling for nanoparticle-based drugs. Nanoinformatics and biomolecular nanomodeling provide a route for effective cancer treatment. Nanoinformatics also enables a data-driven approach to the design of materials to meet health and environmental needs.

Modeling and NanoQSAR

Viewed as a workflow process, nanoinformatics deconstructs experimental studies using data, metadata, controlled vocabularies and ontologies to populate databases so that trends, regularities and theories will be uncovered for use as predictive computational tools. Models are involved at each stage, some material (experiments, reference materials, model organisms) and some abstract (ontology, mathematical formulae), and all intended as a representation of the target system. Models can be used in experimental design, may substitute for experiment or may simulate how a complex system changes over time.

At present, nanoinformatics is an extension of bioinformatics due to the great opportunities for nanotechnology in medical applications, as well as to the importance of regulatory approvals to product commercialization. In these cases, the models target, their purposes, may be physico-chemical, estimating a property based on structure (quantitative structure–property relationship, QSPR); or biological, predicting biological activity based on molecular structure (quantitative structure–activity relationship, QSAR) or the time-course development of a simulation (physiologically based toxicokinetics, PBTK). Each of these has been explored for small molecule drug development with a supporting body of literature.

Particles differ from molecular entities, especially in having surfaces that challenge nomenclature system and QSAR/PBTK model development. For example, particles do not exhibit an octanol–water partition coefficient, which acts as a motive force in QSAR/PBTK models; and they may dissolve in vivo or have band gaps. Illustrative of current QSAR and PBTK models are those of Puzyn and Bachler The OECD has codified regulatory acceptance criteria, and there are guidance roadmaps with supporting workshops to coordinate international efforts.

Communities

Communities active in nanoinformatics include the European Union NanoSafety Cluster, The U.S. National Cancer Institute National Cancer Informatics Program's Nanotechnology Working Group, and the US–EU Nanotechnology Communities of Research.

Individuals who engage in nanoinformatics can be viewed as fitting across four categories of roles and responsibilities for nanoinformatics methods and data:

- Customers, who need either the methods to create the data, the data itself, or both, and who specify the scientific applications and characterization methods and data needs for their intended purposes.
- Creators, who develop relevant and reliable methods and data to meet the needs of customers in the nanotechnology community.
- Curators, who maintain and ensure the quality of the methods and associated data.
- Analysts, who develop and apply methods and models for data analysis and interpretation that are consistent with the quality and quantity of the data and that meet customers' needs.

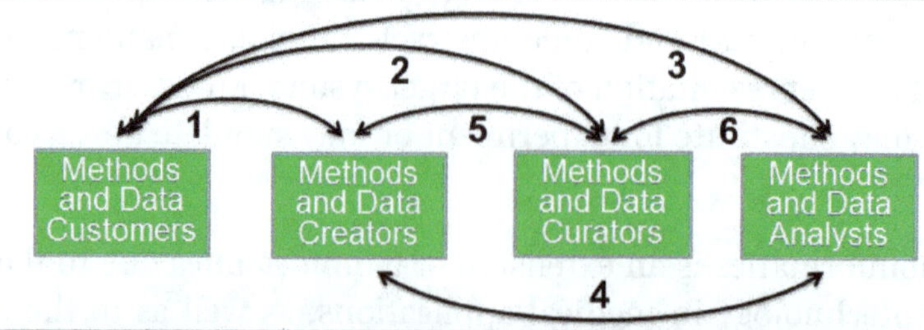

	Set Mission Objectives	Determine Relevance	Collect	Validate	Store	Share	Analyze and Model	Apply	Confirm Effectiveness	Convey Experience	Generalize	Update Guidance
Customers	X	X						X	X	X	X	X
Creators		X	X	X					X			X
Curators		X		X	X	X			X			X
Analysts		X		X			X		X		X	X

Nanoinformatics roles, responsibilities, and communication interfaces.

In some instances, the same individuals perform all four roles. More often, many individuals must interact, with their roles and responsibilities extending over significant distances, organizations, and time. Effective communication is important across each of the twelve links (in both directions across each of the six pairwise interactions) that exist among the various customers, creators, curators, and analysts.

NANOLITHOGRAPHY

Nanolithography is a growing field of techniques within nanotechnology dealing with the engineering (etching, writing, printing) of nanometer-scale structures. " Today, the word has evolved to cover the design of structures in the range of 10^{-9} to 10^{-6} meters, or structures in the nanometer range. Essentially, field is a derivative of lithography, only covering significantly smaller structures. All nanolithographic techniques can be separated into two categories: those that etch away molecules leaving behind the desired structure, and those that directly write the desired structure to a surface (similar to the way a 3D printer creates a structure).

The field of nanolithography was born out of the need to increase the number of transistors in an integrated circuit in order to maintain Moore's Law. While lithographic techniques have been around since the late 18th century, none were applied to nanoscale structures until the mid-1950s. With evolution of the semiconductor industry, demand for techniques capable of producing micro- and nano-scale structures skyrocketed. Photolithography was applied to these structures for the first time in 1958 beginning the age of nanolithography. Since then, photolithography has become the most commercially successful technique, capable of producing sub-100 nm patterns. There are several techniques associated with the field, each designed to serve its many uses in the medical and semiconductor industries. Breakthroughs in this field contribute significantly to the advancement of nanotechnology, and are increasingly important today as demand for smaller and smaller computer chips increases. Further areas of research deal with physical limitations of the field, energy harvesting, and photonics.

Important Techniques

Optical Lithography

Optical Lithography (or photolithography) is one of the most important and prevalent sets of techniques in the nanolithography field. Optical lithography contains several important derivative techniques, all that use very short light wavelengths in order to change the solubility of certain molecules, causing them to wash away in solution, leaving behind a desired structure. Several optical lithography techniques require the use of liquid immersion and a host of resolution enhancement technologies like phase-shift masks (PSM) and optical proximity correction (OPC). Some of the included techniques in this set include multiphoton lithography, X-Ray lithography, light coupling nanolithography (LCM), and extreme ultraviolet lithography (EUVL). This last technique is considered to be the most important next generation lithography (NGL) technique due to its ability to produce structures accurately down below 30 nanometers.

Electron-beam Lithography

Electron beam lithography (EBL) or electron-beam direct-write lithography (EBDW) scans a focused beam of electrons on a surface covered with an electron-sensitive film or resist (e.g. PMMA or HSQ) to draw custom shapes. By changing the solubility of the resist and subsequent selective removal of material by immersion in a solvent, sub-10 nm resolutions have been achieved. This form of direct-write, maskless lithography has high resolution and low throughput, limiting single-column e-beams to photomask fabrication, low-volume production of semiconductor devices, and research and development. Multiple-electron beam approaches have as a goal an increase of throughput for semiconductor mass-production.

EBL can be utilized for selective protein nanopatterning on a solid substrate, aimed for ultrasensitive sensing.

Scanning Probe Lithography

Scanning probe lithography (SPL) is another set of techniques for patterning at the nanometer-scale down to individual atoms using scanning probes, either by etching away unwanted material, or by directly-writing new material onto a substrate. Some of the important techniques in this category include dip-pen nanolithography, thermochemical nanolithography, thermal scanning probe lithography, and local oxidation nanolithography. Dip-pen nanolithography is the most widely used of these techniques.

Nanoimprint Lithography

Nanoimprint lithography (NIL), and its variants, such as Step-and-Flash Imprint Lithography and laser assisted directed imprint (LADI) are promising nanopattern replication technologies where patterns are created by mechanical deformation of imprint resists, typically monomer or polymer formations that are cured by heat or UV light during imprinting. This technique can be combined with contact printing and cold welding. Nanoimprint lithography is capable of producing patterns at sub-10 nm levels.

Quantum Optical Lithography

Quantum optical lithography (QOL), is a diffraction-unlimited method able to write at 1 nm resolution by optical means, using a red laser diode (λ = 650nm).Complex patterns like geometrical figures and letters were obtained at 3 nm resolution on resist substrate. The method was applied to nanopattern graphene at 20 nm resolution.

Miscellaneous Techniques

Stencil lithography is a resist-less and parallel method of fabricating nanometer scale patterns using nanometer-size apertures as shadow-masks.

Magnetolithography

(ML) is based on applying a magnetic field on the substrate using paramagnetic metal masks call "magnetic mask". Magnetic mask which is analog to photomask define the spatial distribution and shape of the applied magnetic field. The second component is ferromagnetic nanoparticles (analog to the Photoresist) that are assembled onto the substrate according to the field induced by the magnetic mask.

Nanosphere Lithography

This technique uses self-assembled monolayers of spheres (typically made of polystyrene) as evaporation masks. This method has been used to fabricate arrays of gold nanodots with precisely controlled spacings.

Proton Beam Writing

This technique uses a focused beam of high energy (MeV) protons to pattern resist material at nanodimensions and has been shown to be capable of producing high-resolution patterning well below the 100 nm mark.

Charged-particle Lithography

This set of techniques include ion- and electron-projection lithographies. Ion beam lithography uses a focused or broad beam of energetic lightweight ions (like He^+) for transferring pattern to a surface. Using Ion Beam Proximity Lithography (IBL) nano-scale features can be transferred on non-planar surfaces.

Neutral Particle Lithography (NPL)

NPL uses a broad beam of energetic neutral particle for pattern transfer on a surface.

NANOMETROLOGY

Metrology is the science of measurements, and nanometrology is that part of metrology that relates to measurements at the nanoscale. Many governments worldwide have existing nanotechnology policies and are taking the preliminary steps towards nanometrology strategies, for example in support of pre-normative R&D and standardization work.

While a first and obvious benefit of metrology is its potential to improve scientific understanding, a second, equally important, but less obvious benefit of metrology is closely linked to the concepts of quality control or conformity assessment, which means making a decision about whether a product or service conforms to specifications.

Conformity assessment provides confidence for the consumer that requirements on products and services are met, it helps producers and suppliers to ensure product quality, and it is essential for reasons of fair trade and of public interest (public health, safety and order, protection of the environment and the consumer). The conformity assessment aspect of metrology turns science and innovation into economy and prosperity. Conformity assessment is especially relevant for nanotechnology, since a great deal of concern exists about the difficulty to turn its scientific developments into innovative products.

"Metrology is necessary as the key discipline to enable the exchange of industrial products or components. To give a few examples, think of the pilot carefully observing his altitude, course, fuel consumption and speed, the food inspectorate measuring bacteria content, environmental engineers measuring air pollution, companies purchasing and selling their product using the same units. Goods and processes can be regulated and legislation can be implemented only because of measurements. Systematic measurement and results with known levels of uncertainty are cornerstones in science and industrial quality control."

Dimensional Nanometrology

The key measurement methods used for dimensional nanometrology are shown using a so-called amplitude–wavelength graph, which plots vertical against lateral dynamic ranges.

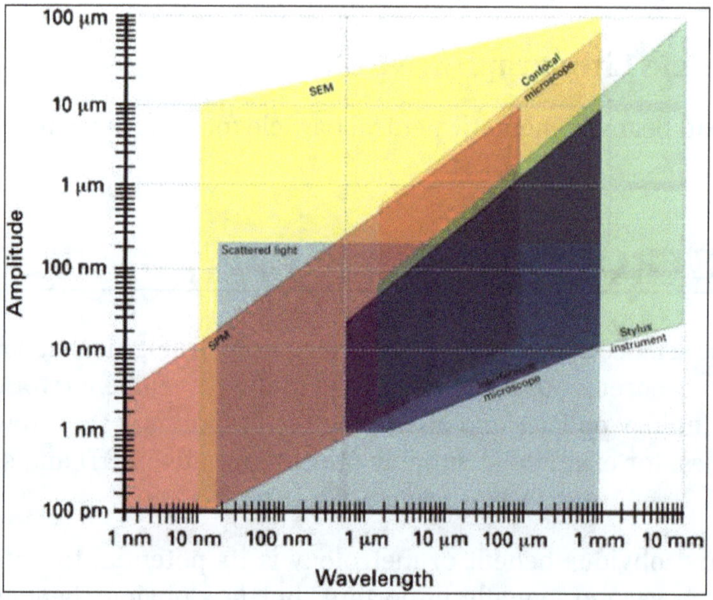

Measurement techniques and their typical dynamic ranges in the lateral (wavelength) and vertical (amplitude) directions.

A recent review covers most of the instruments used in this area. One of the main instrument types used in nanometrology is the scanning probe microscope (SPM), for example the atomic force microscope (AFM).

Chemical Nanometrology

Chemical functionality versus spatial resolution compiled for a range of surface analytical techniques: AES, AFM, dynamic SIMS (dSIMS), electron probe micro-analysis (EPMA), G-SIMS, micro thermal analysis (µTA), scanning near-field optical microscopy (SNOM), scanning tunnelling microscopy (STM), SSIMS, transmission electron microscopy (TEM), parallel electron energy-loss spectrometry (PEELS) and XPS.

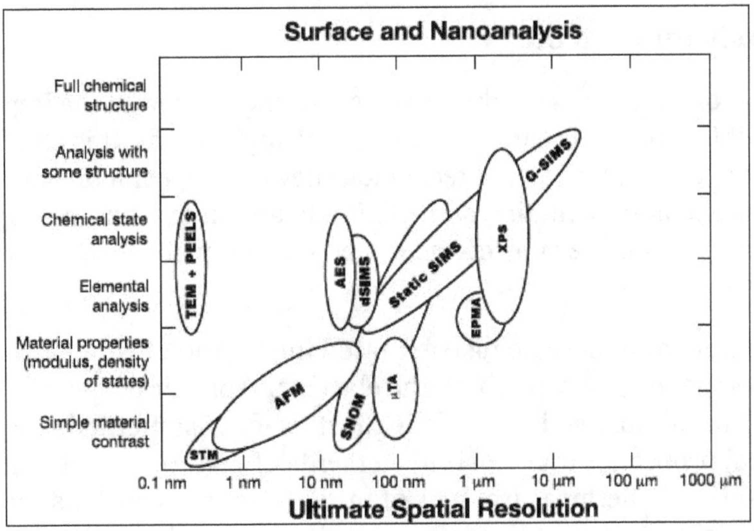

It is important to point out that none of these techniques alone can provide a complete chemical analysis. However, combined information can be given, such as structural properties, composition of chemical species or chemical states contained in the measured sample. Of the techniques shown above, some are primary methods that directly measure an amount of chemical or elemental species, whilst others determine an amount, either after a specific calibration step or expressed as the thickness of the film considered. The traceability of measurements through the mole or the meter is then necessary to provide some metrological analysis.

Thin Film Nanometrology

The applications and economic impact of thin films for nanotechnological products are large, and extend over a wide range of industry sectors. As buried layers, thin films are core to the performance of state-of-art microelectronics and magnetic data storage devices.

As surface functional layers they are employed for wear-, impact- and scratch-resistance, friction control, lubrication, anti-reflection coatings, bioactivity or passivation, wettability, and easy-to-clean or self-cleaning surfaces.

Their importance is not limited to the major end-user industries: The industrial impact of companies specializing in manufacture of thin film deposition equipment and provision of services, as well as in associated metrology and characterization, is significant in its own right.

The generic measurement tasks to be performed in nanometrology of thin films are: Thickness, chemical composition, structure, conformity, uniformity and integrity, surfaces and interfaces (chemistry, structure, roughness, interdiffusion, etc), impurities and dopants, and mechanical properties.

Mechanical Nanometrology

Mechanical nanometrology is a key technology for the support of the important industrial sector of thin films and coatings in industrial applications. It is also important for the development of micro- and nanotechnology devices and components. The most advanced areas are nano-indentation and AFM indentation measurements, with a range of institutes working on the subject and a series of commercial companies selling instruments.

Nano-indentation, or nanoscale instrumented indentation testing is one of the very few techniques that can measure both the elastic and plastic properties of very small volumes of materials and so is one of the most useful test methods for determining the mechanical properties of nanosized elements of materials or coatings. Nano-indentation allows for the measurement of other properties such as modulus, creep and visco-elasticity as well as hardness. Besides applications in material research, nano-indentation is now widely used for quality assurance in industry, mainly in the microelectronics industry and microsystems technology. Nano-indentation is also applied to the investigation and testing of biological matter and for medical diagnosis.

Nanometrology of Structured Materials

Nanostructured materials have internal or surface structure at the nanoscale. Whilst these types of material have been produced for many years, recent advances in manufacturing now mean that nanostructured materials with new properties are being produced. It is important to fully characterize these materials on the nanoscale and build a relationship between this and the macroscale.

The manufacturing route for these materials is either for the nanoscale component to be produced separately from the final material as nano-objects then incorporated into the final structure, or to be produced in situ, for example by lithography.

Examples of such nano-objects include nanoparticles, nanotubes and nanowires. The characterization of materials is an important part of the development process and it serves two broad purposes. Firstly, as part of the research and development into new processes, materials and products, and secondly as quality control and assurance during the pilot plant and manufacturing stages.

There are numerous techniques used to determine the dimensional properties of nanoparticles. However, the techniques can be broadly broken down into three groups. The first group is ensemble techniques where a large number of particles are analyzed simultaneously, for example dynamic light scattering (DLS) and small angle x-ray scattering.

The second group is imaging techniques that analyze individual particles using high-powered microscopic techniques (such as SPM or electron microscopy), where a large amount of information can be obtained on small numbers of particles (typically 200–1000) out of a sample size of many millions.

The third group covers the methods that perform a classification of the nanoparticles in terms of their (apparent) size. Examples are centrifugal liquid sedimentation or field-flow fractionation.

Whilst these techniques are now well established in the development of new materials, their large setup cost and slow speeds mean that they are unsuitable for performing quality control in large manufacturing processes. Bulk quality control will require new standard tests probably based on existing well-established techniques; for example those based on thermal, mechanical, electrical and/or electrochemical properties.

Electrical Nanometrology

Only a small number of metrology institutes have significant work in the field of electrical nanometrology. Electrical nanometrology activities can be divided into the following:

- Support of the metrology for the semiconductor industry. Activities for semiconductor devices have to take into account the International Technology Roadmap for Semiconductors (ITRS).
- Metrology for the electrical quantum triangle related to measurements or improvements on quantum based measurands such as the quantum hall effect, circuits based on Josephson junctions to define voltage over larger ranges and single charge or electron transistors (SETs) to define electric current.
- Investigation into new, or improvements to, methods to reduce the measurement uncertainty and the traceability to support electrical measurements of nanoparticles/aerosols, for example providing traceable electrical current measurements for conducting thin films, nanostructures, organic electronic materials and CNTs.
- Use of nanotechnology to set up improved devices for the quantum Hall effect (QHE) or SETs based on new materials, for example graphene or organic materials.
- Use of the AFM to measure the magnetic properties of nanostructures, for example bits on hard discs.
- Measurements to support spintronics.

Biological Nanometrology

Bionanotechnology is defined as the cross section of life science, healthcare and nanotechnology, which includes the use of nanotechnology to understand and develop biological science and biotechnology, and vice versa.

Bionanotechnology may be considered from two sides: the benefits and risks. There are three main areas where development of bionanotechnology is currently benefiting society, namely (a) medical devices and implants, (b) nanomedicine and personalized medicine, and (c) biological superstructures and engineered bio-nanomaterials. The first two are expected to be of particularly high impact in the next two decades. Bionanotechnology requires interdisciplinary cooperation and novel metrology tools to facilitate its development. There is no single technique applicable for bio-nanometrology. Often a set of techniques is used together for reliable measurements.

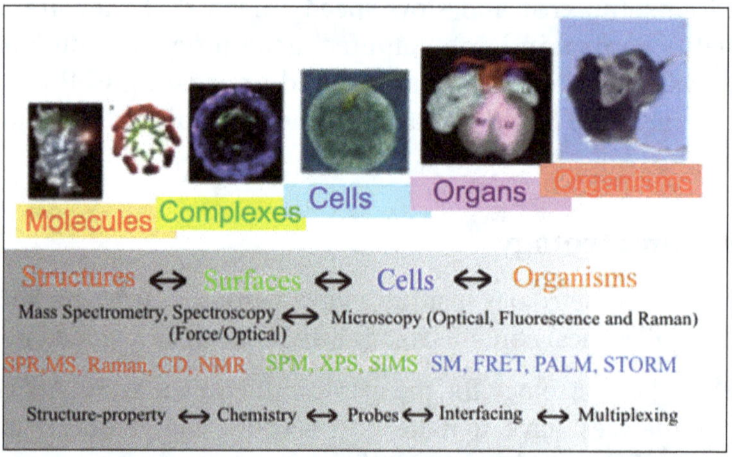

Molecules can be organized in different forms starting from single molecules, complexes and organized in macroscale structures.

NANOFLUIDICS

Schematic diagram of one particular realization of nanofluidics in a nanocapillary array membrane, or NCAM. The NCAM is composed of a large number of parallel nanocapillaries, each of which have a pore radius, $a/2$, which is approximately the same size as the Debye length, κ^{-1}. The electrical double layer is characterized by a counter-ion distribution, N, which is largest at the pore wall and decays toward the center of the pore.

Nanofluidics is the study of the behavior, manipulation, and control of fluids that are confined to structures of nanometer (typically 1–100 nm) characteristic dimensions (1 nm = 10^{-9} m). Fluids confined in these structures exhibit physical behaviors not observed in larger structures, such as those of micrometer dimensions and above, because the characteristic physical scaling lengths of the fluid, (e.g. Debye length, hydrodynamic radius) very closely coincide with the dimensions of the nanostructure itself.

When structures approach the size regime corresponding to molecular scaling lengths, new physical constraints are placed on the behavior of the fluid. For example, these physical constraints induce regions of the fluid to exhibit new properties not observed in bulk, *e.g.* vastly increased viscosity near the pore wall; they may effect changes in thermodynamic properties and may also alter the chemical reactivity of species at the fluid-solid interface. A particularly relevant and useful example is displayed by electrolyte solutions confined in nanopores that contain surface charges, *i.e.* at electrified interfaces, as shown in the nanocapillary array membrane (NCAM) in the accompanying figure.

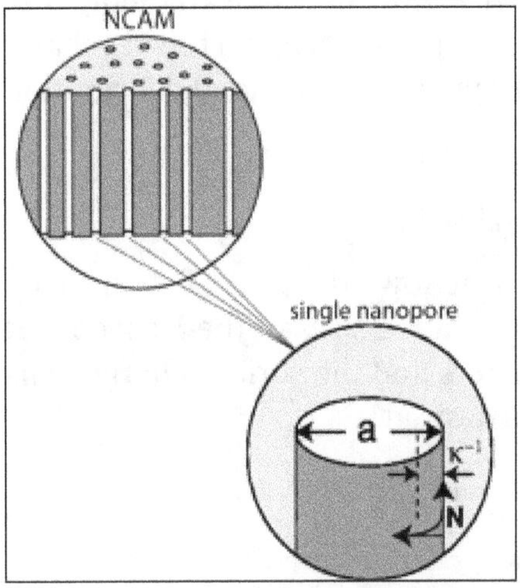

All electrified interfaces induce an organized charge distribution near the surface known as the electrical double layer. In pores of nanometer dimensions the electrical double layer may completely span the width of the nanopore, resulting in dramatic changes in the composition of the fluid and the related properties of fluid motion in the structure. For example, the drastically enhanced surface-to-volume ratio of the pore results in a preponderance of counter-ions (*i.e.* ions charged oppositely to the static wall charges) over co-ions (possessing the same sign as the wall charges), in many cases to the near-complete exclusion of co-ions, such that only one ionic species exists in the pore. This can be used for manipulation of species with selective polarity along the pore length to achieve unusual fluidic manipulation schemes not possible in micrometer and larger structures.

Theory

In 1965, Rice and Whitehead published the seminal contribution to the theory of the transport of electrolyte solutions in long (ideally infinite) nanometer-diameter capillaries. Briefly, the potential, ϕ, at a radial distance, r, is given by the Poisson-Boltzmann equation,

$$\frac{1}{r}\frac{d}{dr}\left(r\frac{d\phi}{dr}\right) = \kappa^2\phi,$$

where κ is the inverse Debye length, $\kappa = \sqrt{\dfrac{8\pi ne^2}{\grave{o}kT}}$, determined by the ion number density, n, the dielectric constant, ε, the Boltzmann constant, k, and the temperature, T. Knowing the potential, $\varphi(r)$, the charge density can then be recovered from the Poisson equation, whose solution may be expressed as a modified Bessel function of the first kind, I_o, and scaled to the capillary radius, a. An equation of motion under combined pressure and electrically-driven flow can then be written:

$$\frac{1}{r}\frac{d}{dr}\left(r\frac{dv_z}{dr}\right) = \frac{1}{\eta}\frac{dp}{dz} - \frac{F_z}{\eta},$$

where η is the viscosity, dp/dz is the pressure gradient, and F_z is the body force driven by the action of the applied electric field, E_z, on the net charge density in the double layer. When there is no applied pressure, the radial distribution of the velocity is given by:

$$v_z(r) = \frac{\grave{o}\phi_o}{4\pi\eta}E_z\left[1 - \frac{I_0(\kappa r)}{I_0(\kappa a)}\right].$$

From the equation above, it follows that fluid flow in nanocapillaries is governed by the κa product, that is, the relative sizes of the Debye length and the pore radius. By adjusting these two parameters and the surface charge density of the nanopores, fluid flow can be manipulated as desired.

Fabrication

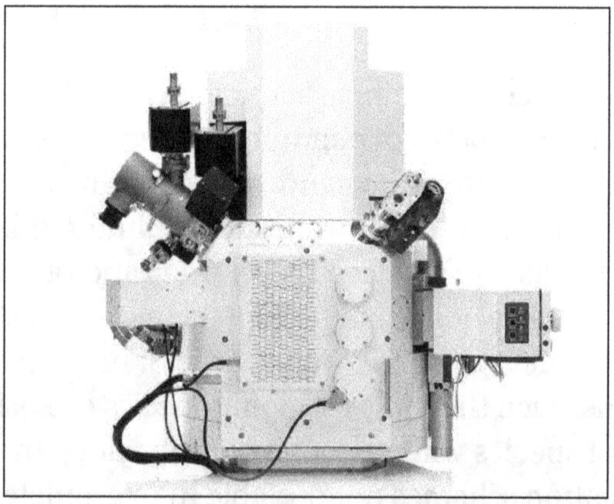

Carl Zeiss Crossbeam 550 - combines a field emission scanning electron microscope (FE-SEM) with a focused ion beam (FIB).

Nanostructures can be fabricated as single cylindrical channels, nanoslits, or nanochannel arrays from materials such as silicon, glass, polymers (e.g. PMMA, PDMS, PCTE) and synthetic vesicles. Standard photolithography, bulk or surface micromachining, replication techniques (embossing, printing, casting and injection molding), and nuclear track or chemical etching, are commonly used to fabricate structures which exhibit characteristic nanofluidic behavior.

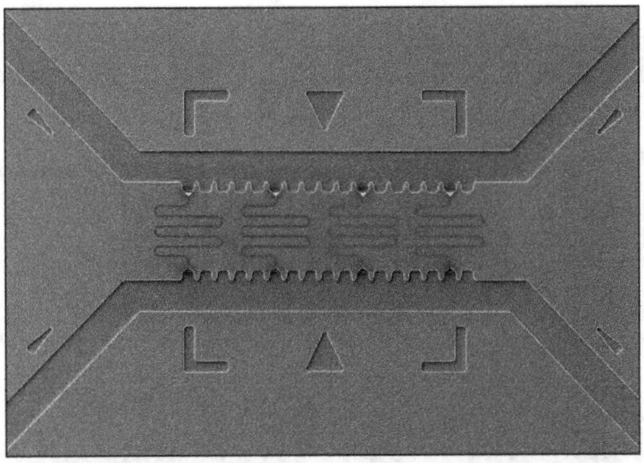

Nanofluidic channels fabricated with a Zeiss Crossbeam 550 L, in a silicon master stamp.

Applications

Because of the small size of the fluidic conduits, nanofluidic structures are naturally applied in situations demanding that samples be handled in exceedingly small quantities, including Coulter counting, analytical separations and determinations of biomolecules, such as proteins and DNA, and facile handling of mass-limited samples. One of the more promising areas of nanofluidics is its potential for integration into microfluidic systems, i.e. micrototal analytical systems or lab-on-a-chip structures. For instance, NCAMs, when incorporated into microfluidic devices, can reproducibly perform digital switching, allowing transfer of fluid from one microfluidic channel to another, selectivity separate and transfer analytes by size and mass, mix reactants efficiently, and separate fluids with disparate characteristics. In addition, there is a natural analogy between the fluid handling capabilities of nanofluidic structures and the ability of electronic components to control the flow of electrons and holes. This analogy has been used to realize active electronic functions such as rectification and field-effect and bipolar transistor action with ionic currents. Application of nanofluidics is also to nano-optics for producing tuneable microlens array.

Nanofluidics have had a significant impact in biotechnology, medicine and clinical diagnostics with the development of lab-on-a-chip devices for PCR and related techniques. Attempts have been made to understand the behaviour of flowfields around nanoparticles in terms of fluid forces as a function of Reynolds and Knudsen number using computational fluid dynamics. The relationship between lift, drag and Reynolds number has been shown to differ dramatically at the nanoscale compared with macroscale fluid dynamics.

Challenges

There are a variety of challenges associated with the flow of liquids through carbon nanotubes and nanopipes. A common occurrence is channel blocking due to large macromolecules in the liquid. Also, any insoluble debris in the liquid can easily clog the tube. A solution for this researchers are hoping to find is a low friction coating or channel materials that help reduce the blocking of the tubes. Also, large polymers, including biologically relevant molecules such as DNA, often fold *in vivo,* causing blockages.

Typical DNA molecules from a virus have lengths of approx. 100–200 kilobases and will form a random coil of the radius some 700 nm in aqueous solution at 20%. This is also several times greater than the pore diameter of even large carbon pipes and two orders of magnitude the diameter of a single walled carbon nanotube.

NANOSCALE TECHNOLOGY

Nanoscale technology is a branch of nanotechnology in which standard size tools are used to manufacture simple structures and devices with dimensions on the order of a few nanometer s or less, where one nanometer (1 nm) is equal to a billionth of a meter (10^{-9} m). Nanoscale technology encompasses all of nanotechnology except molecular manufacturing , which involves the use of nanoscale (extremely small) tools to build structures, devices, and systems at the molecular level.

Nanotechnology has potential benefits in a number of fields, including water purification, sanitation, agriculture, alternative energy (particularly photovoltaics), home and business construction, computer manufacturing, communications, and medicine. Nanotechnology is also considered to pose risks, including economic disruption, misuse by people or organizations with nefarious intent, evolution of extreme governmental regulation, controversies over morality, and damage to the environment.

References

- Molecular-nanotechnology-mnt: circuitstoday.com, Retrieved 05 March, 2020
- Ehud Gazit, Plenty of room for biology at the bottom: An introduction to bionanotechnology. Imperial College Press, 2007, ISBN 978-1-86094-677-6
- Minary-Jolandan, Majid; Yu, Min-Feng (2009). "Nanomechanical heterogeneity in the gap and overlap regions of type I collagen fibrils with implications for bone heterogeneity". Biomacromolecules. 10 (9): 2565–70. doi:10.1021/bm900519v. PMID 19694448
- Katsuhiko Ariga, Michihiro Nishikawa, Taizo Mori, Jun Takeya, Lok Kumar Shrestha & Jonathan P. Hill (2019). "Self-assembly as a key player for materials nanoarchitectonics". Science and Technology of Advanced Materials. 20 (1): 51–95. doi:10.1080/14686996.2018.1553108. PMC 6374972. PMID 30787960
- Diana, De la Iglesia; Stacey, Harper; Mark D, Hoover; Fred, Klaessig; Phil, Lippell; Bettye, Maddux; Jeffrey, Morse; Andre, Nel; Krishna, Rajan; Rebecca, Reznik-Zellen; Mark T., Tuominen (2011). "Nanoinformatics 2020 Roadmap" (PDF). National Nanomanufacturing Network: 9–13. doi:10.4053/rp001-110413
- Esmek, Franziska M.; Bayat, Parisa; Pérez-Willard, Fabián; Volkenandt, Tobias; Blick, Robert H.; Fernandez-Cuesta, Irene (2019). "Sculpturing wafer-scale nanofluidic devices for DNA single molecule analysis". Nanoscale. 11 (28): 13620–13631. doi:10.1039/C9NR02979F. ISSN 2040-3364
- Nanoscale-technology: whatis.techtarget.com, Retrieved 25 June, 2020

3

Nanomaterials

The materials whose size lies between 1 and 100 nanometers are known as nanomaterials. Some of these are nanocomposites, silicon and carbon nanotubes, nanoshells, nanofibers, silicene, nanosheets and nanomesh. The topics elaborated in this chapter will help in gaining a better perspective of nanomaterials.

Nanomaterials are any type of material of nanosized thickness, i.e. less than 100 nm in thickness. There are various types, many of which exhibit different properties than bulk materials. One common factor of nanomaterials is that this thickness range is also known as the quantum regime, where quantum effects play a major role in defining the properties. Because of this, nanomaterials often fall into different dimensional categories, be it 2D, 1D or 0D.

Some people define the nanomaterial range as 1-1000 nm; however, the widely accepted range is between 1 and 100 nm. This range is also the quantum regime and is why many nanomaterials exhibit quantum effects that make them suitable for electronics, semiconductor technologies and quantum technologies. So, even though there are many different classes of nanomaterials, including thin films, coatings and nanoparticles, the focus of this article is to introduce the various nanomaterials by the dimension that they exist in.

Aside from the specific thickness range, 50% of the particles of the nanomaterial must be present in the 1-100 nm range and not be aggregated for materials to be classed as nanomaterials. For layered materials or single walled carbon nanotubes (SWCNT), one or more external dimensions need to be less than 1 nm thickness (i.e., Angstrom thickness).

Nanomaterials exist in different dimensions, not only because they can be one atomic layer thick, but by how their electrons can be confined to flow in a certain number of dimensions. For example 2D materials have their electrons confined in one direction, so the electrons then move in two directions, hence the name. The same principle applies for 1D and 0D materials which have their electrons confined in 2 and 3 dimensions respectively; and their electrons can move in 1 and 0 directions respectively. So, let's look at some examples.

2D Materials

Uniatomic 2D Materials

There are many types of uniatomic 2D materials, such as germanene (made from germanium atoms), stanene (tin), silicene (silicon), phosphorene (phosphorous) and, of course, graphene (carbon).

Graphene is by far the most useful and the closest material to commercialization within this list, especially as some of these are still theoretical materials. However, graphene and the various other 2D atomic materials possess an excellent array of optical, physical and electrical properties that make them useful for a wide range of applications. Once graphene has been successfully used across many applications at a commercial level, it is expected that many of the other 2D materials will follow suit, although it could take a while.

MXenes

Outside of graphene, the class of MXene 2D materials show some of the best electronic properties. The most common MXene is boron nitride, which exhibits a hexagonal array of alternating boron and nitrogen atoms. Many people think that the MXenes show better properties than graphene, but they are much harder to synthesize. As such, they have not been as widely studied as graphene. However interest in them is significantly growing.

Whilst boron nitride is the most common, and the most widely researched, MXenes come in many forms and are often made from a combination of early transition metals (M), such as Titanium, Vanadium, Chromium and Niobium, alongside carbon or nitrogen (X). Future applications involving the MXenes could include EMI shielding, water purification and in energy storage systems.

TMDCs

Transition metal dichalcogenides (TMDCs) are one of the oldest and longest studied class of 2D materials. They are widely used in semiconducting and electronic applications, but do not have as wide a range in properties in each material as other 2D materials. However, there are over 100 TMDCs to date, with the most common being tungsten diselenide (WSe_2), tungsten disulphide (WS_2), molybdenum disulphide (MoS_2) and molybdenum diselenide ($MoSe_2$), although many other transition metals and chalcogen atoms can be used.

One characteristic feature of TMDCs, is that one monolayer is composed of three atomic layers, where a layer of metal atoms is sandwiched between two layers of chalcogen atoms (note that these layers are physically bonded and are not held by van der Waals forces).

1D Materials

Nanotubes

Nanotubes, be it a carbon nanotube or inorganic nanotube, are materials which are elongated in one dimension, with a length-to-diameter ratio of up to 132,000,000:1. Nanotubes direct electrons along the elongated axis and come in many forms, including single-walled nanotubes (SWNT), multi-walled nanotubes (MWNT), chiral nanotubes, armchair nanotubes and zigzag nanotubes.

There has been a lot of hype about how carbon nanotubes could be used for many applications, especially in structural applications. However, issues with dispersing and aligning carbon nanotubes led to them to go out of favour for a while. They have recently been making a resurgence as many of these issues have been negated.

Nanowires

Nanowires, otherwise known as quantum wires, are another well-known 1D material. Again, nanowires are elongated in one direction, albeit with a much lower width to length ratio of 1:1000. The most common nanowires are silver nanowires, which are also known to be highly electrically conductive. Nanowires are known for exhibiting many different quantum effects, which alongside their unidirectional electron movement, have made them ideal materials for various electronic applications.

0D Materials

Quantum Dots

Quantum dots are a very common and useful type of nanoparticle, where the electrons are confined in all 3-dimensions. Quantum dots are small semiconducting particles that have been greatly used in displays and solar cells. Quantum dots emit a certain wavelength of light when they encounter either light or electricity and many quantum dots can be easily tuned. Quantum dots composed of cadmium, such as cadmium selenide (CdSe), are the widest class of quantum dots that have been studied.

Fullerenes

Fullerenes, also termed Buckminsterfullerenes (because of the geodesic shape they exhibit) come in two forms - pure carbon fullerenes and endohedral fullerenes, with the difference being that endohedral fullerenes contain additional atom(s) inside the carbon fullerene.

Fullerenes come in many shapes and sizes, but the most common is C_{60} as it is the most energetically and structurally stable form. Fullerenes composed of boron have also been predicted. Carbon fullerenes are composed of both single and double bonds, which are arranged into pentagons and hexagons. It is the pentagons which give the fullerenes their curvature. All fullerenes contain 12 pentagons, with a differing number of hexagons.

Nanoparticles

Overall, nanoparticles come in many forms. There are too many to individually discuss, but some of the most common are: single element nanoparticles, such as silver and gold nanoparticles which are used in medical imaging; metal oxide nanoparticles, including titanium dioxide nanoparticles used in white paint formulations; and amphiphilic nanoparticles such as Janus particles, which are used as stabilizers. Janus particles are an interesting class of nanoparticle, as one half of the surface is different to the other, and these two surfaces can differ in their external receptors, hydrophobicity or hydrophilicity, surface charge, and even in their magnetism.

NANOCOMPOSITE

Nanocomposite is a multiphase solid material where one of the phases has one, two or three dimensions of less than 100 nanometers (nm) or structures having nano-scale repeat distances between the different phases that make up the material.

The idea behind Nanocomposite is to use building blocks with dimensions in nanometre range to design and create new materials with unprecedented flexibility and improvement in their physical properties.

In the broadest sense this definition can include porous media, colloids, gels and copolymers, but is more usually taken to mean the solid combination of a bulk matrix and nano-dimensional phase(s) differing in properties due to dissimilarities in structure and chemistry. The mechanical, electrical, thermal, optical, electrochemical, catalytic properties of the nanocomposite will differ markedly from that of the component materials. Size limits for these effects have been proposed:

- <5 nm for catalytic activity.
- <20 nm for making a hard magnetic material soft.

- <50 nm for refractive index changes.
- <100 nm for achieving superparamagnetism, mechanical strengthening or restricting matrix dislocation movement.

Nanocomposites are found in nature, for example in the structure of the abalone shell and bone. The use of nanoparticle-rich materials long predates the understanding of the physical and chemical nature of these materials. Jose-Yacaman investigated the origin of the depth of colour and the resistance to acids and bio-corrosion of Maya blue paint, attributing it to a nanoparticle mechanism. From the mid-1950s nanoscale organo-clays have been used to control flow of polymer solutions (e.g., as paint viscosifiers) or the constitution of gels (e.g., as a thickening substance in cosmetics, keeping the preparations in homogeneous form).

In mechanical terms, nanocomposites differ from conventional composite materials due to the exceptionally high surface to volume ratio of the reinforcing phase and/or its exceptionally high aspect ratio. The reinforcing material can be made up of particles (e.g. minerals), sheets (e.g. exfoliated clay stacks) or fibres (e.g. carbon nanotubes or electrospun fibres). The area of the interface between the matrix and reinforcement phase(s) is typically an order of magnitude greater than for conventional composite materials. The matrix material properties are significantly affected in the vicinity of the reinforcement. With polymer nanocomposites, properties related to local chemistry, degree of thermoset cure, polymer chain mobility, polymer chain conformation, degree of polymer chain ordering or crystallinity can all vary significantly and continuously from the interface with the reinforcement into the bulk of the matrix.

This large amount of reinforcement surface area means that a relatively small amount of nanoscale reinforcement can have an observable effect on the macroscale properties of the composite. For example, adding carbon nanotubes improves the electrical and thermal conductivity. Other kinds of nanoparticulates may result in enhanced optical properties, dielectric properties, heat resistance or mechanical properties such as stiffness, strength and resistance to wear and damage. In general, the nano reinforcement is dispersed into the matrix during processing. The percentage by weight (called mass fraction) of the nanoparticulates introduced can remain very low (on the order of 0.5% to 5%) due to the low filler percolation threshold, especially for the most commonly used non-spherical, high aspect ratio fillers (e.g., nanometer-thin platelets, such as clays, or nanometer-diameter cylinders, such as carbon nanotubes). The orientation and arrangement of asymmetric nanoparticles, thermal property mismatch at the interface, interface density per unit volume of nanocomposite, and polydispersity of nanoparticles significantly affect the effective thermal conductivity of nanocomposites.

Ceramic-matrix Nanocomposites

Ceramic matrix composites (CMCs) consist of ceramic fibers embedded in a ceramic matrix. The matrix and fibers can consist of any ceramic material, including carbon and carbon fibers. The ceramic occupying most of the volume is often from the group of oxides, such as nitrides, borides, silicides, whereas the second component is often a metal. Ideally both components are finely dispersed in each other in order to elicit particular optical, electrical and magnetic properties as well as tribological, corrosion-resistance and other protective properties.

The binary phase diagram of the mixture should be considered in designing ceramic-metal nanocomposites and measures have to be taken to avoid a chemical reaction between both components. The last point mainly is of importance for the metallic component that may easily react with the ceramic and thereby lose its metallic character. This is not an easily obeyed constraint because the preparation of the ceramic component generally requires high process temperatures. The safest measure thus is to carefully choose immiscible metal and ceramic phases. A good example of such a combination is represented by the ceramic-metal composite of TiO_2 and Cu, the mixtures of which were found immiscible over large areas in the Gibbs' triangle of ' Cu-O-Ti.

The concept of ceramic-matrix nanocomposites was also applied to thin films that are solid layers of a few nm to some tens of μm thickness deposited upon an underlying substrate and that play an important role in the functionalization of technical surfaces. Gas flow sputtering by the hollow cathode technique turned out as a rather effective technique for the preparation of nanocomposite layers. The process operates as a vacuum-based deposition technique and is associated with high deposition rates up to some μm/s and the growth of nanoparticles in the gas phase. Nanocomposite layers in the ceramics range of composition were prepared from TiO_2 and Cu by the hollow cathode technique that showed a high mechanical hardness, small coefficients of friction and a high resistance to corrosion.

Metal-matrix Nanocomposites

Metal matrix nanocomposites can also be defined as reinforced metal matrix composites. This type of composites can be classified as continuous and non-continuous reinforced materials. One of the more important nanocomposites is Carbon nanotube metal matrix composites, which is an emerging new material that is being developed to take advantage of the high tensile strength and electrical conductivity of carbon nanotube materials. Critical to the realization of CNT-MMC possessing optimal properties in these areas are the development of synthetic techniques that are (a) economically producible, (b) provide for a homogeneous dispersion of nanotubes in the metallic matrix, and (c) lead to strong interfacial adhesion between the metallic matrix and the carbon nanotubes. In addition to carbon nanotube metal matrix composites, boron nitride reinforced metal matrix composites and carbon nitride metal matrix composites are the new research areas on metal matrix nanocomposites.

A recent study, comparing the mechanical properties (Young's modulus, compressive yield strength, flexural modulus and flexural yield strength) of single- and multi-walled reinforced polymeric (polypropylene fumarate—PPF) nanocomposites to tungsten disulfide nanotubes reinforced PPF nanocomposites suggest that tungsten disulfide nanotubes reinforced PPF nanocomposites possess significantly higher mechanical properties and tungsten disulfide nanotubes are better reinforcing agents than carbon nanotubes. Increases in the mechanical properties can be attributed to a uniform dispersion of inorganic nanotubes in the polymer matrix (compared to carbon nanotubes that exist as micron sized aggregates) and increased crosslinking density of the polymer in the presence of tungsten disulfide nanotubes (increase in crosslinking density leads to an increase in the mechanical properties). These results suggest that inorganic nanomaterials, in general, may be better reinforcing agents compared to carbon nanotubes.

Another kind of nanocomposite is the energetic nanocomposite, generally as a hybrid sol–gel with a silica base, which, when combined with metal oxides and nano-scale aluminum powder, can form superthermite materials.

Polymer-matrix Nanocomposites

In the simplest case, appropriately adding nanoparticulates to a polymer matrix can enhance its performance, often dramatically, by simply capitalizing on the nature and properties of the nanoscale filler (these materials are better described by the term nanofilled polymer composites). This strategy is particularly effective in yielding high performance composites, when uniform dispersion of the filler is achieved and the properties of the nanoscale filler are substantially different or better than those of the matrix. The uniformity of the dispersion is in all nanocomposites is counteracted by thermodynamically driven phase separation. Clustering of nanoscale fillers produces aggregates that serve as structural defects and result in failure. Layer-by-layer (LbL) assembly when nanometer scale layers of nanoparticulates and a polymers are added one by one. LbL composites display performance parameters 10-1000 times better that the traditional nanocomposites made by extrusion or batch-mixing.

Nanoparticles such as graphene, carbon nanotubes, molybdenum disulfide and tungsten disulfide are being used as reinforcing agents to fabricate mechanically strong biodegradable polymeric nanocomposites for bone tissue engineering applications. The addition of these nanoparticles in the polymer matrix at low concentrations (~0.2 weight %) cause significant improvements in the compressive and flexural mechanical properties of polymeric nanocomposites. Potentially, these nanocomposites may be used as a novel, mechanically strong, light weight composite as bone implants. The results suggest that mechanical reinforcement is dependent on the nanostructure morphology, defects, dispersion of nanomaterials in the polymer matrix, and the cross-linking density of the polymer. In general, two-dimensional nanostructures can reinforce the polymer better than one-dimensional nanostructures, and inorganic nanomaterials are better reinforcing agents than carbon based nanomaterials. In addition to mechanical properties, polymer nanocomposites based on carbon nanotubes or graphene have been used to enhance a wide range of properties, giving rise to functional materials for a wide range of high added value applications in fields such as energy conversion and storage, sensing and biomedical tissue engineering. For example, multi-walled carbon nanotubes based polymer nanocomposites have been used for the enhancement of the electrical conductivity.

Nanoscale dispersion of filler or controlled nanostructures in the composite can introduce new physical properties and novel behaviors that are absent in the unfilled matrices. This effectively changes the nature of the original matrix (such composite materials can be better described by the term genuine nanocomposites or hybrids). Some examples of such new properties are fire resistance or flame retardancy, and accelerated biodegradability.

A range of polymeric nanocomposites are used for biomedical applications such as tissue engineering, drug delivery, cellular therapies. Due to unique interactions between polymer and nanoparticles, a range of property combinations can be engineered to mimic native tissue structure and properties. A range of natural and synthetic polymers are used to design polymeric nanocomposites for biomedical applications including starch, cellulose, alginate, chitosan, collagen, gelatin, and fibrin, poly(vinyl alcohol) (PVA), poly(ethylene glycol) (PEG), poly(caprolactone) (PCL), poly(lactic-co-glycolic acid) (PLGA), and poly(glycerol sebacate) (PGS). A range of nanoparticles including ceramic, polymeric, metal oxide and carbon-based nanomaterials are incorporated within polymeric network to obtain desired property combinations.

Magnetic Nanocomposites

Nanocomposites that can respond to an external stimulus are of increased interest due to the fact that, due to the large amount of interaction between the phase interfaces, the stimulus response can have a larger effect on the composite as a whole. The external stimulus can take many forms, such as a magnetic, electrical, or mechanical field. Specifically, magnetic nanocomposites are useful for use in these applications due to the nature of magnetic material's ability to respond both to electrical and magnetic stimuli. The penetration depth of a magnetic field is also high, leading to an increased area that the nanocomposite is affected by and therefore an increased response. In order to respond to a magnetic field, a matrix can be easily loaded with nanoparticles or nanorods The different morphologies for magnetic nanocomposite materials are vast, including matrix dispersed nanoparticles, core-shell nanoparticles, colloidal crystals, macroscale spheres, or janus-type nanostructures.

Magnetic nanocomposites can be utilized in a vast number of applications, including catalytic, medical, and technical. For example, palladium is a common transition metal used in catalysis reactions. Magnetic nanoparticle-supported palladium complexes can be used in catalysis to increase the efficiency of the palladium in the reaction.

Magnetic nanocomposites can also be utilized in the medical field, with magnetic nanorods embedded in a polymer matrix can aid in more precise drug delivery and release. Finally, magnetic nanocomposites can be used in high frequency/high temperature applications. For example, multi-layer structures can be fabricated for use in electronic applications. An electrodeposited Fe/Fe oxide multi-layered sample can be an example of this application of magnetic nanocomposites.

ENGINEERING CONTROLS FOR NANOMATERIALS

Engineering controls for nanomaterials are a set of hazard control methods and equipment for workers who interact with nanomaterials. Engineering controls are physical changes to the workplace that isolate workers from hazards, and are considered the most important set of methods for controlling the health and safety hazards of nanomaterials after systems and facilities have been designed.

A fume hood is an example of an engineering control that uses local exhaust ventilation combined with an enclosure to isolate a worker from nanomaterials that may become airborne.

The primary hazard of nanomaterials is health effects from inhalation of aerosols containing nanoparticles. Many engineering controls developed for other industries can be used or adapted for protecting workers from exposure to nanomaterials, including ventilation and filtering using laboratory fixtures such as fume hoods, containment using gloveboxes, and other non-ventilation controls such as sticky mats. Research is ongoing as to what engineering controls are most effective for nanomaterials.

Engineering Controls

Controlling exposures to occupational hazards is considered the fundamental method of protecting workers. Traditionally, a hierarchy of controls has been used as a means of determining how to implement feasible and effective controls, which typically include elimination, substitution, engineering controls, administrative controls, and personal protective equipment. Methods earlier in the list are considered generally more effective in reducing the risk associated with a hazard, with process changes and engineering controls recommended as the primary means for reducing exposures, and personal protective equipment being the approach of last resort. Following the hierarchy is intended to lead to the implementation of inherently safer systems, ones where the risk of illness or injury has been substantially reduced.

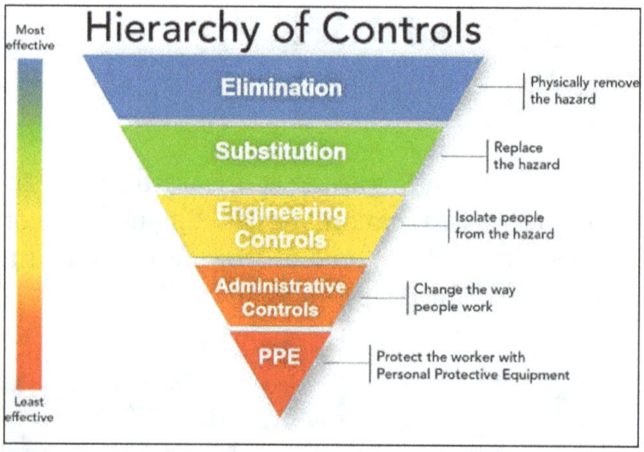

Engineering controls are the third most effective member of the hierarchy of hazard controls. They are preferred over administrative controls and personal protective equipment, but are less preferred than elimination or substitution of the hazards.

Engineering controls are physical changes to the workplace that isolate workers from hazards by containing them in an enclosure, or removing contaminated air from the workplace through ventilation and filtering. Well-designed engineering controls are typically passive, in the sense of being independent of worker interactions, which reduces the potential for worker behavior to impact exposure levels. They also ideally do not interfere with productivity and ease of processing for the worker, because otherwise the operator may be motivated to circumvent the controls. The initial cost of engineering controls can be higher than administrative controls or personal protective equipment, but the long-term operating costs are frequently lower, and can sometimes provide cost savings in other areas of the process.

Nanomaterials

Nanomaterials have at least one primary dimension of less than 100 nanometers, and often have properties different from those of their bulk components that are technologically useful. Because nanotechnology is a recent development, the health and safety effects of exposures to nanomaterials, and what levels of exposure may be acceptable, is not yet fully understood. Processing and manufacturing of nanomaterials involve a wide range of hazards. The types of engineering controls optimal for each situation is influenced by the quantity and dustiness of the material as well as the duration of the task. For example, stronger engineering controls are indicated if dry nanomaterials cannot be substituted with a suspension, or if procedures such as sonication or cutting of a solid matrix containing nanomaterials cannot be eliminated.

As with any new technology, the earliest exposures are expected to occur among workers conducting research in laboratories and pilot plants. It is recommended that researchers handling engineered nanomaterials in these contexts perform that work in a manner that is protective of their safety and health. Control measures for nanoparticles, dusts, and other hazards are most effective when implemented within the context of a comprehensive occupational safety and health management system, the critical elements of which include management commitment and employee involvement, worksite analysis, hazard prevention and control, and sufficient training for employees, supervisors, and managers.

Ventilation

Ventilation systems are distinguished as being either local or general. Local exhaust ventilation operates at or near the source of contamination, often in conjunction with an enclosure, while general exhaust ventilation operates on an entire room through a building's HVAC system.

Local Exhaust Ventilation

Local exhaust ventilation (LEV) is the application of an exhaust system at or near the source of contamination. If properly designed, it will be much more efficient at removing contaminants than dilution ventilation, requiring lower exhaust volumes, less make-up air, and, in many cases, lower costs. By applying exhaust at the source, contaminants are removed before they get into the general work environment. Examples of local exhaust systems include fume hoods, vented balance enclosures, and biosafety cabinets. Exhaust hoods lacking an enclosure are less preferable, and laminar flow hoods are not recommended because they direct air outwards towards the worker.

In 2006, a survey was conducted of international nanotechnology firms and research laboratories that reported manufacturing, handling, researching, or using nanomaterials. All organizations participating in the survey reported using some type of engineering control. The most common exposure control used was the traditional laboratory fume hood, with two-thirds of firms reporting its use.

Fume Hoods

Fume hoods are recommended to have an average inward velocity of 80–100 feet per minute (fpm) at the face of the hood. For higher toxicity materials, a higher face velocity of 100–120 fpm is recommended in order to provide better protection. However, face velocities exceeding 150 fpm are not believed to improve performance, and could increase hood leakage.

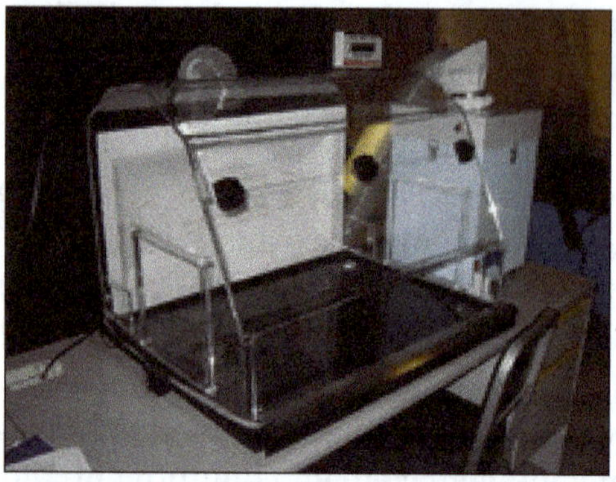

Vented balance enclosures used in the pharmaceutical industry can be used for nanomaterials, with the advantages of smaller size and lower turbulence.

New fume hoods specifically designed for nanotechnology are being developed primarily based on low-turbulence balance enclosures, which were initially developed for the weighing of pharmaceutical powders; these nanomaterial handling enclosures provide adequate containment at lower face velocities, typically operating at 65–85 fpm. They are useful for weighing operations, which disturb the nanomaterial and increase its aerosolization.

It is recommended that air exiting a fume hood should be passed through a HEPA filter and exhausted outside the work environment, with used filters being handled as hazardous waste. Turbulence can cause nanomaterials to exit the front of the hood, and can be avoided by keeping the sash in the proper position, keeping the interior of the hood uncluttered with equipment, and not making fast movements while working. High face velocities can result in loss of powdered nanomaterials; while as of 2012 there was little research on the effectiveness for low-flow fume hoods, there was evidence that air curtain hoods were effective at containing nanoparticles.

Other Enclosures

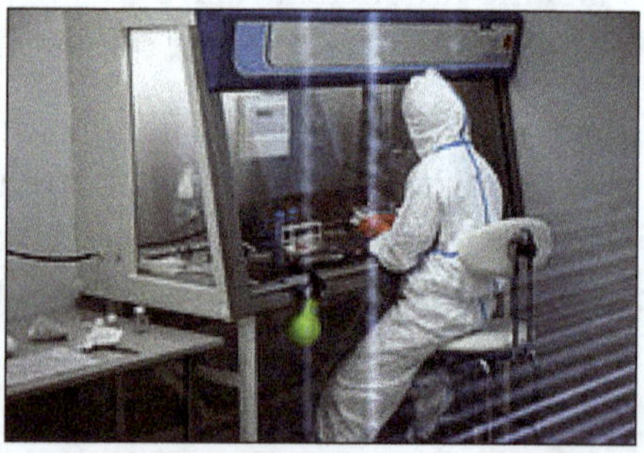

Biosafety cabinets, although designed to contain bioaerosols, can also be used to contain nanomaterials.

Biosafety cabinets are designed to contain bioaerosols, which have a similar size to engineered nanoparticles, and are believed to be effective with nanoparticles. However, common biosafety cabinets are more prone to turbulence. As with fume hoods, they are recommended to be exhausted outside the facility.

Dedicated large-scale ventilated enclosures for large pieces of equipment can also be used.

General Exhaust Ventilation

General exhaust ventilation (GEV), also called dilution ventilation, is different from local exhaust ventilation because instead of capturing emissions at their source and removing them from the air, general exhaust ventilation allows the contaminant to be emitted into the workplace air and then dilutes the concentration of the contaminant to an acceptable level. GEV is inefficient and costly as compared to local exhaust ventilation, and given the lack of established exposure limits for most nanomaterials, they are not recommended to be relied upon for controlling exposure.

However, GEV can provide negative room pressure to prevent contaminants from exiting the room. The use of supply and exhaust air throughout the facility can provide pressurization schemes that reduce the number of workers exposed to potentially hazardous materials, for example keeping production areas at a negative pressure with respect to nearby areas. For general exhaust ventilation in laboratories, a nonrecirculating system is used with 4–12 air changes per hour when used in tandem with local exhaust ventilation, and sources of contamination are placed close to the air exhaust and downwind of workers, and away from windows or doors that may cause air drafts.

Control Verification

Several control verification techniques can be used to assess room airflow patterns and verify the proper operation of LEV systems. It is considered important to confirm that an LEV system is operating as designed by regularly measuring exhaust airflows. A standard measurement, hood static pressure, provides information on airflow changes that affect hood performance. For hoods designed to prevent exposures to hazardous airborne contaminants, the American Conference of Governmental Industrial Hygienists recommends the installation of a fixed hood static pressure gauge.

Additionally, Pitot tubes, hot-wire anemometers, smoke generators, and dry ice tests can be used to qualitatively measure hood slot/face and duct air velocity, while tracer-gas leak testing is a quantitative method. Standardized testing and certification procedures such as ANSI Z9.5 and ASHRAE 110 can be used, as can qualitative indicators of proper installation and functionality such as inspection of gaskets and hoses.

Containment

Containment refers to the physical isolation of a process or a piece of equipment to prevent the release of the hazardous material into the workplace. It can be used in conjunction with ventilation measures to provide an enhanced level of protection for nanomaterial workers. Examples include placing equipment that may release nanomaterials in a separate room. Standard dust control methods such as enclosures for conveyor systems or using a sealed system for bag filling are effective at reducing respirable dust concentrations.

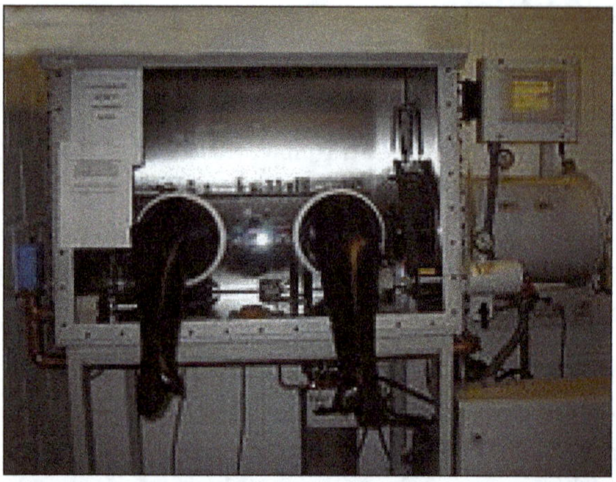

Gloveboxes are completely enclosed, but are more difficult to use than fume hoods, and may leak if used under positive pressure.

Non-ventilation engineering controls can also include devices developed for the pharmaceutical industry, including isolation containment systems. One of the most common flexible isolation systems is glovebox containment, which can be used as an enclosure around small-scale powder processes, such as mixing and drying. Rigid glovebox isolation units also provide a method for isolating the worker from the process and are often used for medium-scale operations involving transfer of powders. Glovebags are similar to rigid gloveboxes, but they are flexible and disposable. They are used for small operations for containment or protection from contamination. Gloveboxes are sealed systems that provide a high degree of operator protection, but are more difficult to use due to limited mobility and size of operation. Transferring materials into and out of the enclosure also is an exposure risk. In addition, some gloveboxes are configured to use positive pressure, which can increase the risk of leaks.

Another non-ventilation control used in this industry is the continuous liner system, which allows the filling of product containers while enclosing the material in a polypropylene bag. This system is often used for off-loading materials when the powders are to be packed into drums.

Other Engineering Controls

Other non-ventilation engineering controls in general cover a range of control measures, such as guards and barricades, material treatment, or additives. One example is placing walk-off sticky mats at room exits. Antistatic devices can be used when handling nanomaterials to reduce their electrostatic charge, making them less likely to disperse or adhere to clothing. Water spray application is also an effective method for reducing respirable dust concentrations.

A sticky mat in a nanomaterials production facility. Ideally, other engineering controls should lessen the amount of dust collecting on the floor and being tracked onto the sticky mat, unlike this example.

NANOMATERIAL-BASED CATALYST

Nanomaterial-based catalysts are usually heterogeneous catalysts broken up into metal nanoparticles in order to enhance the catalytic process. Metal nanoparticles have high surface area, which can increase catalytic activity. Nanoparticle catalysts can be easily separated and recycled. They are typically used under mild conditions to prevent decomposition of the nanoparticles.

Functionalized Nanoparticles

Functionalized metal nanoparticles are more stable toward solvents compared to non-functionalized metal nanoparticles. In liquids, the metal nanoparticles can be affected by van der Waals force. Particle aggregation can sometimes decrease catalytic activity by lowering the surface area. Nanoparticles can also be functionalized with polymers or oligomers to sterically stabilize the nanoparticles by providing a protective layer that prevents the nanoparticles from interacting with each other. Alloys of two metals, called bimetallic nanoparticles, are used to create synergistic effects on catalysis between the two metals.

Potential Applications

Dehalogenation and Hydrogenation

Nanoparticle catalysts are active for the hydrogenolysis of C-Cl bonds such as polychlorinated biphenyls. Another reaction is hydrogenation of halogenated aromatic amines is also important for the synthesis of herbicides and pesticides as well as diesel fuel. In organic chemistry, hydrogenation of a C-Cl bond with deuterium is used to selectively label the aromatic ring for use in experiments dealing with the kinetic isotope effect. Buil created rhodium complexes that generated rhodium nanoparticles. These nanoparticles catalyzed the dehalogenation of aromatic compounds as well as the hydrogenation of benzene to cyclohexane. Polymer-stabilized nanoparticles can also be used for the hydrogenation of cinnamaldehyde and citronellal. Yu found that the ruthenium nanocatalysts are more selective in the hydrogenation of citronellal compared to the traditional catalysts used.

Hydrosilylation Reactions

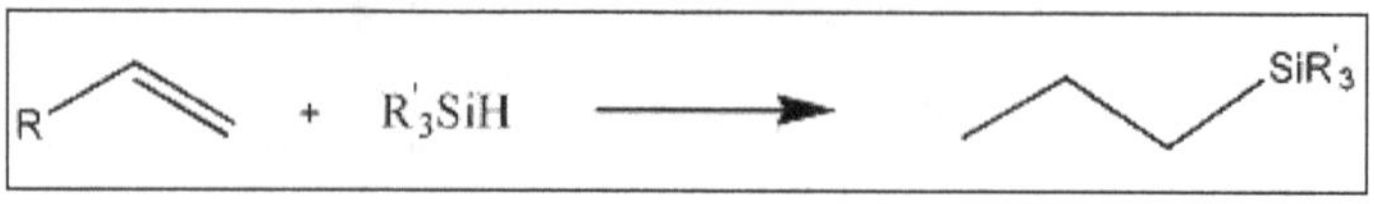

Hydrosilylation reaction.

The reduction of gold, cobalt, nickel, palladium, or platinum organometallic complexes with silanes produces metal nanoparticle that catalyze the hydrosilylation reaction. BINAP-functionalized palladium nanoparticles and gold nanoparticles have been used for the hydrosilylaytion of styrene under mild conditions; they were found to be more catalytically active and more stable than non-nanoparticle Pd-BINAP complexes. The reaction may also be catalyzed by a nanoparticle that consists of two metals.

Organic Redox Reactions

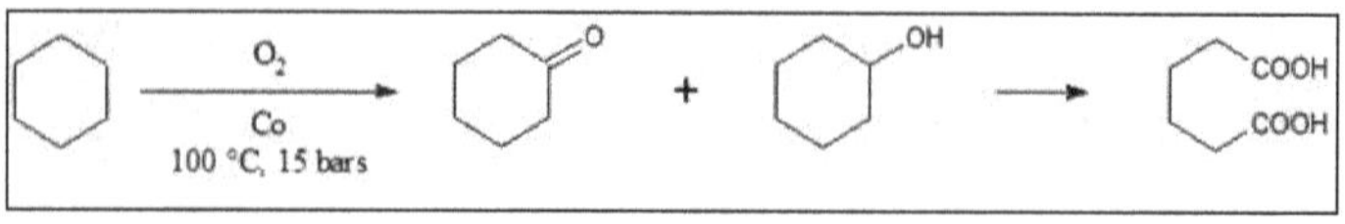

Oxidation reaction of cyclohexane to synthesize adiapic acid.

An oxidation reaction to form adipic acid is shown in figure and it can be catalyzed by cobalt nanoparticles. This is used in an industrial scale to produce the nylon 6,6 polymer. Other examples of oxidation reactions that are catalyzed by metallic nanoparticles include the oxidation of cyclooctane, the oxidation of ethene, and glucose oxidation.

C-C Coupling Reactions

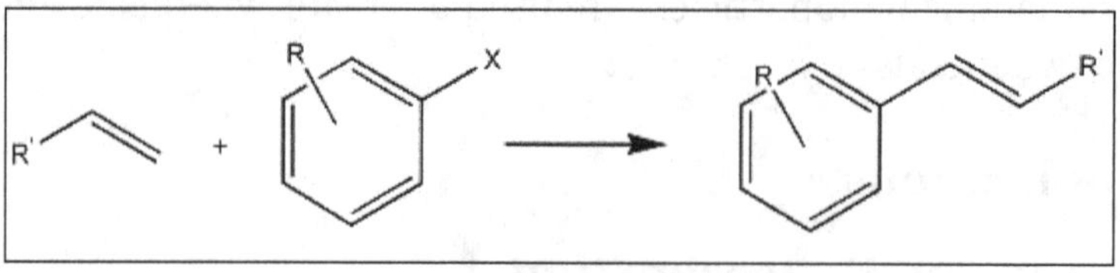

Heck coupling reaction.

Metallic nanoparticles can catalyze C–C coupling reactions such as the hydroformylation of olefins, the synthesis of vitamin E and the Heck coupling and Suzuki coupling reactions.

Palladium nanoparticles were found to efficiently catalyze Heck coupling reactions. It was found that increased electronegativity of the ligands on the palladium nanoparticles increased their catalytic activity.

The compound $Pd_2(dba)_3$ is a source of Pd(0), which is the catalytically active source of palladium used for many reactions, including cross coupling reactions. Pd2(dba)3 was thought to be a homogeneous catalytic precursor, but recent articles suggest that palladium nanoparticles are formed, making it a heterogeneous catalytic precursor.

Alternative Fuels

Iron oxide and cobalt nanoparticles can be loaded onto various surface active materials like alumina to convert gases such as carbon monoxide and hydrogen into liquid hydrocarbon fuels using the Fischer-Tropsch process.

Much research on nanomaterial-based catalysts has to do with maximizing the effectiveness of the catalyst coating in fuel cells. Platinum is currently the most common catalyst for this application, however, it is expensive and rare, so a lot of research has been going into maximizing the catalytic properties of other metals by shrinking them to nanoparticles in the hope that someday they will be an efficient and economic alternative to platinum. Gold nanoparticles also exhibit catalytic properties, despite the fact that bulk gold is unreactive.

Yttrium stabilized zirconium nanoparticles were found to increase the efficiency and reliability of a solid oxide fuel cell. Nanomaterial ruthenium/platinum catalysts could potentially be used to catalyze the purification of hydrogen for hydrogen storage. Palladium nanoparticles can be functionalized with organometallic ligands to catalyze the oxidation of CO and NO to control air pollution in the environment. Carbon nanotube supported catalysts can be used as a cathode catalytic support for fuel cells and metal nanoparticles have been used to catalyze the growth of carbon nanotubes. Platinum-cobalt bimetallic nanoparticles combined with carbon nanotubes are promising candidates for direct methanol fuel cells since they produce a higher stable current electrode.

Medicine

In magnetic chemistry, nanoparticles can be used for catalyst support for medicinal use.

Nanozymes

Besides conventional catalysis, nanomaterials have been explored for mimicking natural enzymes. The nanomaterials with enzyme mimicking activities are termed as nanozymes. Lots of nanomaterials have been used to mimic varieties of natural enzymes, such as oxidase, peroxidase, catalase, SOD, nuclease, etc. The nanozymes have found wide applications in many areas, from biosensing and bioimaging to therapeutics and water treatment.

Nanostructures for Electrocatalysis

Nanocatalysts are of wide interest in fuel cells and electrolyzers, where the catalyst strongly affects efficiency.

Nanoporous Surfaces

In fuel cells, nanoporous materials are widely used to make cathodes. Porous nanoparticles of platinum have good activity in nanocatalysis but are less stable and their lifetime is short.

Nanoparticles

One drawback to the use of nanoparticles is their tendency to agglomerate. The problem can be mitigated with the correct catalyst support. Nanoparticles are optimal structures to be used as nanosensors because they can be tuned to detect specific molecules. Examples of Pd nanoparticles electrodeposited on multi-walled carbon nanotubes have shown good activity towards catalysis of cross-coupling reactions.

Nanowires

Nanowires are very interesting for electrocatalytic purpose because they are easier to produce and the control over their characteristics in the production process is quite precise. Also, nanowires can increase faradaic efficiency due to their spatial extent and thus to greater availability of reactants on the active surface.

Materials

The nanostructures involved in electrocatalysis processes can be made up of different materials. Through the use of nanostructured materials, electrocatalysts can achieve good physical-chemical stability, high activity, good conductivity and low cost. Metallic nanomaterials are commonly made up of transition metals (mostly iron, cobalt, nickel, palladium, platinum). Multi-metal nanomaterials show new properties due to the characteristics of each metal. The advantages are the increase in activity, selectivity and stability and the cost reduction. Metals can be combined in different ways such as in the core-shell bimetallic structure: the cheapest metal forms the core and the most active one (typically a noble metal) constitutes the shell. By adopting this design, the use of rare and expensive metals can be reduced down to 20%.

One of the future challenges is to find new stable materials, with good activity and especially low cost. Metallic glasses, polymeric carbon nitride (PCN) and materials derived from metal-organic frameworks (MOF) are just a few examples of materials with electrocatalytic properties on which research is currently investing.

Photocatalysis

Many of the photocatalytic systems can benefit from the coupling with a noble metal; the first Fujishima-Honda cell made use of a co-catalyst plate as well. For instance, the essential design of a disperse photocatalytic reactor for water splitting is that of a water sol in which the dispersed phase is made up of semiconductor quantum dots each coupled to a metallic co-catalyst: the QD converts the incoming electromagnetic radiation into an exciton whilst the co-catalyst acts as an electron scavenger and lowers the overpotential of the electrochemical reaction.

Characterization of Nanoparticles

Some techniques that can be used to characterize functionalized nanomaterial catalysts include X-ray photoelectron spectroscopy, transmission electron microscopy, circular dichroism spectroscopy, nuclear magnetic resonance spectroscopy, UV-visible spectroscopy and related experiments.

NANOPARTICLE

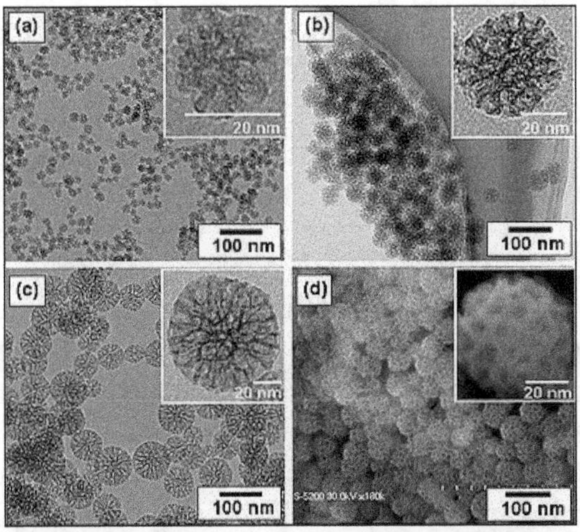

TEM (a, b, and c) images of prepared mesoporous silica nanoparticles with mean outer diameter: (a) 20nm, (b) 45nm, and (c) 80nm. SEM (d) image corresponding to (b). The insets are a high magnification of mesoporous silica particle.

Nanoparticle are particles between 1 and 100 nanometres (nm) in size with a surrounding interfacial layer. The interfacial layer is an integral part of nanoscale matter, fundamentally affecting all of its properties. The interfacial layer typically consists of ions, inorganic and organic molecules. Organic molecules coating inorganic nanoparticles are known as stabilizers, capping and surface ligands, or passivating agents. In nanotechnology, a particle is defined as a small object that behaves as a whole unit with respect to its transport and properties. Particles are further classified according to diameter.

According to ISO Technical Specification 80004, a nanoparticle is defined as a nano-object with all three external dimensions in the nanoscale, whose longest and shortest axes do not differ significantly, with a significant difference typically being a factor of at least 3. Nanoparticles can exhibit size-related properties significantly different from those of either fine particles or bulk materials.

Nanoclusters have at least one dimension between 1 and 10 nanometers and a narrow size distribution. Nanopowders are agglomerates of ultrafine particles, nanoparticles, or nanoclusters. Nanometer-sized single crystals, or single-domain ultrafine particles, are often referred to as nanocrystals.

"Ultrafine particles" (UFP) are synonymous with nanoparticles and range between 1 and 100 nm in size, as opposed to "fine particles" sized between 100 and 2,500 nm and "coarse particles" ranging from 2,500 to 10,000 nm. During the 1970s and 80s, when the first thorough fundamental studies with nanoparticles were underway in the United States (by Granqvist and Buhrman) and Japan (within an ERATO Project), researchers used the term ultrafine particles. However, during the 1990s, before the National Nanotechnology Initiative was launched in the United States, the term nanoparticle had become more common.

The terms colloid and nanoparticle are not interchangeable. A colloid is a mixture which has solid particles dispersed in a liquid medium. The term applies only if the particles are larger than atomic dimensions but small enough to exhibit Brownian motion, with the critical size range (or particle diameter) typically ranging from nanometers (10^{-9} m) to micrometers (10^{-6} m). Colloids can contain particles too large to be nanoparticles, and nanoparticles can exist in non-colloidal form, for examples as a powder or in a solid matrix.

Although nanoparticles are associated with modern science, they have a long history. Nanoparticles were used by artisans as far back as Rome in the fourth century in the famous Lycurgus cup made of dichroic glass as well as the ninth century in Mesopotamia for creating a glittering effect on the surface of pots. In modern times, pottery from the Middle Ages and Renaissance often retains a distinct gold- or copper-colored metallic glitter. This luster is caused by a metallic film that was applied to the transparent surface of a glazing. The luster can still be visible if the film has resisted atmospheric oxidation and other weathering.

The luster originates within the film itself, which contains silver and copper nanoparticles dispersed homogeneously in the glassy matrix of the ceramic glaze. These nanoparticles are created by the artisans by adding copper and silver salts and oxides together with vinegar, ochre, and clay on the surface of previously-glazed pottery. The object is then placed into a kiln and heated to about 600 °C in a reducing atmosphere. In heat the glaze softens, causing the copper and silver ions to migrate into the outer layers of the glaze. There the reducing atmosphere reduced the ions back to metals, which then came together forming the nanoparticles that give the color and optical effects. Luster technique showed that ancient craftsmen had a sophisticated empirical knowledge of materials. The technique originated in the Muslim world. As Muslims were not allowed to use gold in artistic representations, they sought a way to create a similar effect without using real gold. The solution they found was using luster.

Michael Faraday provided the first description, in scientific terms, of the optical properties of nanometer-scale metals in his classic 1857 paper. In a subsequent paper, the author (Turner) points out that: "It is well known that when thin leaves of gold or silver are mounted upon glass and heated to a temperature that is well below a red heat (~500 °C), a remarkable change of properties takes place, whereby the continuity of the metallic film is destroyed. The result is that white light is now freely transmitted, reflection is correspondingly diminished, while the electrical resistivity is enormously increased."

Properties

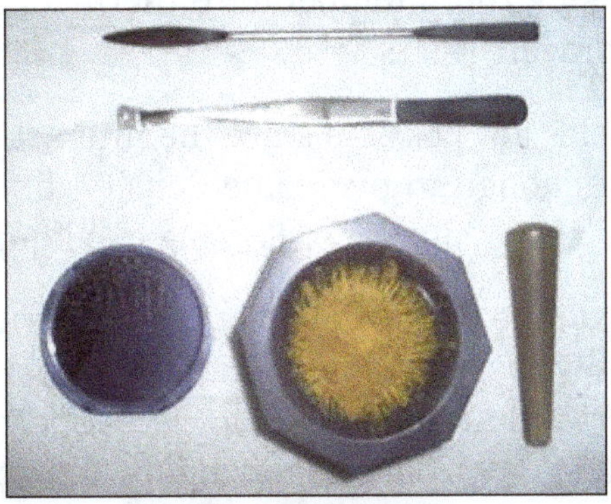

Silicon nanopowder.

1 kg of particles of 1 mm^3 has the same surface area as 1 mg of particles of 1 nm^3.

Nanoparticles are of great scientific interest as they are, in effect, a bridge between bulk materials and atomic or molecular structures. A bulk material should have constant physical properties regardless of its size, but at the nano-scale, size-dependent properties are often observed. Thus, the properties of materials change as their size approaches the nanoscale and as the percentage of the surface in relation to the percentage of the volume of a material becomes significant. For bulk materials larger than one micrometer (or micron), the percentage of the surface is insignificant in relation to the volume in the bulk of the material. The interesting and sometimes unexpected properties of nanoparticles are therefore largely due to the large surface area of the material, which dominates the contributions made by the small bulk of the material.

Nanoparticles often possess unexpected physical and optical properties, as they are small enough to confine their electrons and produce quantum effects. For example, gold nanoparticles appear deep-red to black in solution. Nanoparticles of yellow gold and grey silicon are red in color. Gold nanoparticles melt at much lower temperatures (~300 °C for 2.5 nm size) than bulk gold (1064 °C). Absorption of solar radiation is much higher in materials composed of nanoparticles than in thin films of continuous sheets of material. In both solar PV and solar thermal applications, by controlling the size, shape, and material of the particles, it is possible to control solar absorption.

Other size-dependent property changes include quantum confinement in semiconductor particles, localized surface plasmon in some metal particles and superparamagnetism in magnetic materials. Some changes in physical properties are not always desirable; for example, ferromagnetic materials smaller than 10 nm can switch their magnetisation direction using room temperature thermal energy, thus making them unsuitable for memory storage.

Suspensions of nanoparticles are possible since the interaction of the particle surface with the solvent is strong enough to overcome density differences, which otherwise usually result in a material either sinking or floating in a liquid.

The high surface area to volume ratio of nanoparticles provides a tremendous driving force for diffusion, especially at elevated temperatures. Sintering can take place at lower temperatures, over shorter time scales than for larger particles. In theory, this does not affect the density of the final product, though flow difficulties and the tendency of nanoparticles to agglomerate complicates matters. Moreover, nanoparticles have been found to impart some extra properties to various day to day products. For example, the presence of titanium dioxide nanoparticles imparts what is known as the self-cleaning effect, which lend useful water-repellant and antibacterial properties to paints and other products. Zinc oxide nanoparticles have been found to have superior UV blocking properties and are widely used in the preparation of sunscreen lotions, being completely photostable though toxic.

Clay nanoparticles when incorporated into polymer matrices increase reinforcement, leading to stronger plastics, verifiable by a higher glass transition temperature and other mechanical property tests. These nanoparticles are hard, and impart their properties to the polymer (plastic). Nanoparticles have also been attached to textile fibers in order to create smart and functionWal clothing.

Metal, dielectric, and semiconductor nanoparticles have been formed, as well as hybrid structures (e.g., core–shell nanoparticles). The core-shell nanoparticles can support simultaneously both electric and magnetic resonances, demoonstrating entirely new properties when compared with bare metallic nanoparticles if the resonances are properly engineered. The formation of the core-shell structure from two different metals enables an energy exchange between the core and the shell, typically found in upconverting nanoparticles and downconverting nanoparticles, and causes a shift in the emission wavelength spectrum. By introducing a dielectric layer, plasmonic core (metal)-shell (dielectric) nanoparticles enhance light absorption by increasing scattering. Recently, the metal core-dielectric shell nanoparticle has demonstrated a zero backward scattering with enhanced forward scattering on a silicon substrate when surface plasmon is located in front of a solar cell.

Nanoparticles made of semiconducting material may also be labeled quantum dots if they are small enough (typically sub 10 nm) that quantization of electronic energy levels occurs. Such nanoscale particles are used in biomedical applications as drug carriers or imaging agents with work being done to try to understand the fluid dynamic properties (e.g. drag forces) in nanoscale applications. This has shown the relationship between the fluid forces on nanoparticles and the fluid Reynolds and Knudsen numbers.

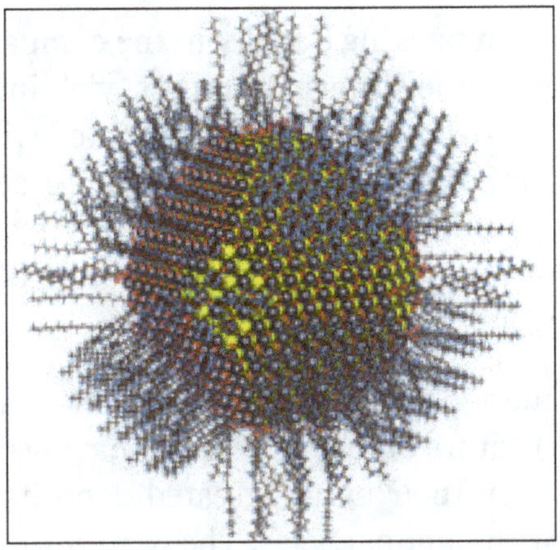

Semiconductor nanoparticle (quantum dot) of lead sulfide with complete passivation by oleic acid, oleyl amine and hydroxyl ligands (size ~5nm).

Semi-solid and soft nanoparticles have been produced. A prototype nanoparticle of semi-solid nature is the liposome. Various types of liposome nanoparticles are currently used clinically as delivery systems for anticancer drugs and vaccines.

Nanoparticles with one half hydrophilic and the other half hydrophobic are termed Janus particles and are particularly effective for stabilizing emulsions. They can self-assemble at water/oil interfaces and act as solid surfactants.

Hydrogel nanoparticles made of N-isopropylacrylamide hydrogel core shell can be dyed with affinity baits, internally. These affinity baits allow the nanoparticles to isolate and remove undesirable proteins while enhancing the target analytes.

Variation in Properties

The chemical processing and synthesis of high-performance technological components for the private, industrial, and military sectors requires the use of high-purity ceramics (oxide ceramics, such as aluminium oxide or copper(II) oxide), polymers, glass-ceramics, and composite materials, as metal carbides (SiC), nitrides (Aluminum nitrides, Silicon nitride), metals (Al, Cu), non-metals (graphite, carbon nanotubes) and layered (Al + Aluminium carbonate, Cu + C). In condensed bodies formed from fine powders, the irregular particle sizes and shapes in a typical powder often lead to non-uniform packing morphologies that result in packing density variations in the powder compact.

Uncontrolled agglomeration of powders due to attractive van der Waals forces can also give rise to microstructural heterogeneity. Differential stresses that develop as a result of non-uniform drying shrinkage are directly related to the rate at which the solvent can be removed, and thus highly dependent upon the distribution of porosity. Such stresses have been associated with a plastic-to-brittle transition in consolidated bodies, and can yield to crack propagation in the unfired body if not relieved.

In addition, any fluctuations in packing density in the compact as it is prepared for the kiln are often amplified during the sintering process, yielding inhomogeneous densification. Some pores and other structural defects associated with density variations have been shown to play a detrimental role in the sintering process by growing and thus limiting end-point densities. Differential stresses arising from inhomogeneous densification have also been shown to result in the propagation of internal cracks, thus becoming the strength-controlling flaws.

Inert gas evaporation and inert gas deposition are free many of these defects due to the distillation (cf. purification) nature of the process and having enough time to form single crystal particles, however even their non-aggreated deposits have lognormal size distribution, which is typical with nanoparticles. The reason why modern gas evaporation techniques can produce a relatively narrow size distribution is that aggregation can be avoided. However, even in this case, random residence times in the growth zone, due to the combination of drift and diffusion, result in a size distribution appearing lognormal.

It would, therefore, appear desirable to process a material in such a way that it is physically uniform with regard to the distribution of components and porosity, rather than using particle size distributions that will maximize the green density. The containment of a uniformly dispersed assembly of strongly interacting particles in suspension requires total control over interparticle forces. Monodisperse nanoparticles and colloids provide this potential.

Monodisperse powders of colloidal silica, for example, may therefore be stabilized sufficiently to ensure a high degree of order in the colloidal crystal or polycrystalline colloidal solid that results from aggregation. The degree of order appears to be limited by the time and space allowed for longer-range correlations to be established. Such defective polycrystalline colloidal structures would appear to be the basic elements of submicrometer colloidal materials science and, therefore, provide the first step in developing a more rigorous understanding of the mechanisms involved in microstructural evolution in high performance materials and components.

Synthesis

There are several methods for creating nanoparticles, including gas condensation, attrition, chemical precipitation, ion implantation, pyrolysis and hydrothermal synthesis. In attrition, macro- or micro-scale particles are ground in a ball mill, a planetary ball mill, or other size-reducing mechanism. The resulting particles are air classified to recover nanoparticles. In pyrolysis, a vaporous precursor (liquid or gas) is forced through an orifice at high pressure and burned. The resulting solid (a version of soot) is air classified to recover oxide particles from by-product gases. Traditional pyrolysis often results in aggregates and agglomerates rather than single primary particles. Ultrasonic nozzle spray pyrolysis (USP) on the other hand aids in preventing agglomerates from forming.

A thermal plasma can deliver the energy to vaporize small micrometer-size particles. The thermal plasma temperatures are in the order of 10,000 K, so that solid powder easily evaporates. Nanoparticles are formed upon cooling while exiting the plasma region. The main types of the thermal plasma torches used to produce nanoparticles are dc plasma jet, dc arc plasma, and radio frequency (RF) induction plasmas. In the arc plasma reactors, the energy necessary for evaporation and reaction is provided by an electric arc formed between the anode and the cathode. For example, silica sand can be vaporized with an arc plasma at atmospheric pressure, or thin aluminum wires can be vaporized by exploding wire method. The resulting mixture of plasma gas and silica vapour can be rapidly cooled by quenching with oxygen, thus ensuring the quality of the fumed silica produced.

In RF induction plasma torches, energy coupling to the plasma is accomplished through the electromagnetic field generated by the induction coil. The plasma gas does not come in contact with electrodes, thus eliminating possible sources of contamination and allowing the operation of such plasma torches with a wide range of gases including inert, reducing, oxidizing, and other corrosive atmospheres. The working frequency is typically between 200 kHz and 40 MHz. Laboratory units run at power levels in the order of 30–50 kW, whereas the large-scale industrial units have been tested at power levels up to 1 MW. As the residence time of the injected feed droplets in the plasma is very short, it is important that the droplet sizes are small enough in order to obtain complete evaporation. The RF plasma method has been used to synthesize different nanoparticle materials, for example synthesis of various ceramic nanoparticles such as oxides, carbours/carbides, and nitrides of Ti and Si.

Inert-gas Condensation

Inert-gas condensation is frequently used to produce metallic nanoparticles. The metal is evaporated in a vacuum chamber containing a reduced atmosphere of an inert gas. Condensation of the supersaturated metal vapor results in creation of nanometer-size particles, which can be entrained in the inert gas stream and deposited on a substrate or studied in situ. Early studies were based on thermal evaporation. Using magnetron sputtering to create the metal vapor allows to achieve higher yields. The method can easily be generalized to alloy nanoparticles by choosing appropriate metallic targets.

The use of sequential growth schemes, where the particles travel through a second metallic vapor, results in growth of core-shell (CS) structures.

Radiolysis Method

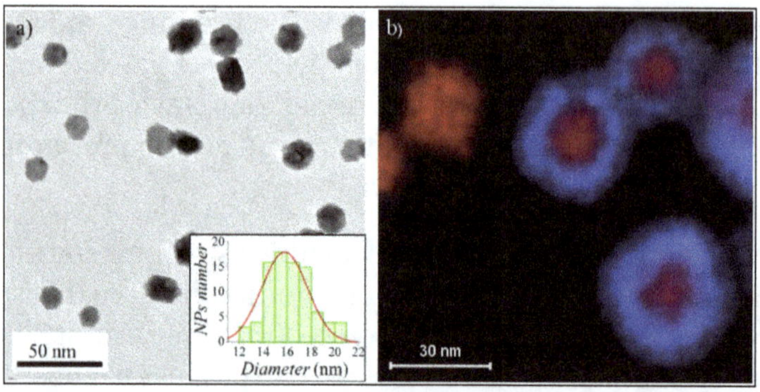

a) Transmission electron microscopy (TEM) image of Hf nanoparticles grown by magnetron-sputtering inert-gas condensation (inset: size distribution) and b) energy dispersive x-ray (EDX) mapping of Ni and Ni@Cu core@shell nanoparticles.

Nanoparticles can also be formed using radiation chemistry. Radiolysis from gamma rays can create strongly active free radicals in solution. This relatively simple technique uses a minimum number of chemicals. These including water, a soluble metallic salt, a radical scavenger (often a secondary alcohol), and a surfactant (organic capping agent). High gamma doses on the order of 10^4 Gray are required. In this process, reducing radicals will drop metallic ions down to the zero-valence state. A scavenger chemical will preferentially interact with oxidizing radicals to prevent the re-oxidation of the metal. Once in the zero-valence state, metal atoms begin to coalesce into particles. A chemical surfactant surrounds the particle during formation and regulates its growth. In sufficient concentrations, the surfactant molecules stay attached to the particle. This prevents it from dissociating or forming clusters with other particles. Formation of nanoparticles using the radiolysis method allows for tailoring of particle size and shape by adjusting precursor concentrations and gamma dose.

Sol–gel

The sol–gel process is a wet-chemical technique (also known as chemical solution deposition) widely used recently in the fields of materials science and ceramic engineering. Such methods are used primarily for the fabrication of materials (typically a metal oxide) starting from a chemical solution (*sol*, short for solution), which acts as the precursor for an integrated network (or *gel*) of either discrete particles or network polymers.

Typical precursors are metal alkoxides and metal chlorides, which undergo hydrolysis and polycondensation reactions to form either a network "elastic solid" or a colloidal suspension (or dispersion) – a system composed of discrete (often amorphous) submicrometer particles dispersed to various degrees in a host fluid. Formation of a metal oxide involves connecting the metal centers with oxo (M-O-M) or hydroxo (M-OH-M) bridges, therefore generating metal-oxo or metal-hydroxo polymers in solution. Thus, the sol evolves toward the formation of a gel-like diphasic system containing both a liquid phase and solid phase whose morphologies range from discrete particles to continuous polymer networks.

In the case of the colloid, the volume fraction of particles (or particle density) may be so low that a significant amount of fluid may need to be removed initially for the gel-like properties to be recognized. This can be accomplished in a number of ways. The simplest method is to allow time for sedimentation to occur, and then pour off the remaining liquid. Centrifugation can also be used to accelerate the process of phase separation.

Removal of the remaining liquid (solvent) phase requires a drying process, which typically causes shrinkage and densification. The rate at which the solvent can be removed is ultimately determined by the distribution of porosity in the gel. The ultimate microstructure of the final component will clearly be strongly influenced by changes implemented during this phase of processing. Afterward, a thermal treatment, or firing process, is often necessary in order to favor further polycondensation and enhance mechanical properties and structural stability via final sintering, densification, and grain growth. One of the distinct advantages of using this methodology as opposed to the more traditional processing techniques is that densification is often achieved at a much lower temperature.

The precursor sol can be either deposited on a substrate to form a film (e.g., by dip-coating or spin-coating), cast into a suitable container with the desired shape (e.g., to obtain a monolithic ceramics, glasses, fibers, membranes, aerogels), or used to synthesize powders (e.g., microspheres, nanospheres). The sol–gel approach is a cheap and low-temperature technique that allows for the fine control of the product's chemical composition. Even small quantities of dopants, such as organic dyes and rare earth metals, can be introduced in the sol and end up uniformly dispersed in the final product. It can be used in ceramics processing and manufacturing as an investment casting material, or as a means of producing very thin films of metal oxides for various purposes. Sol–gel derived materials have diverse applications in optics, electronics, energy, space, (bio)sensors, medicine (e.g., controlled drug release) and separation (e.g., chromatography) technology.

Ion Implantation

Ion implantation may be used to treat the surfaces of dielectric materials such as sapphire and silica to make composites with near-surface dispersions of metal or oxide nanoparticles.

Morphology

Scientists have taken to naming their particles after the real-world shapes that they might represent. Nanospheres, nanorods, nanochains, nanoreefs, nanoboxes and more have appeared in the literature. These morphologies sometimes arise spontaneously as an effect of a templating or directing agent present in the synthesis such as miscellar emulsions or anodized alumina pores, or from the innate crystallographic growth patterns of the materials themselves. Some of these morphologies may serve a purpose, such as long carbon nanotubes used to bridge an electrical junction, or just a scientific curiosity like the stars shown at right.

Nanostars of vanadium(IV) oxide.

Amorphous particles usually adopt a spherical shape (due to their microstructural isotropy), whereas the shape of anisotropic microcrystalline whiskers corresponds to their particular crystal habit. At the small end of the size range, nanoparticles are often referred to as clusters. Spheres, rods, fibers, and cups are just a few of the shapes that have been grown. The study of fine particles is called micromeritics.

Characterization

Nanoparticles have different analytical requirements than conventional chemicals, for which chemical composition and concentration are sufficient metrics. Nanoparticles have other physical properties that must be measured for a complete description, such as size, shape, surface properties, crystallinity, and dispersion state. Additionally, sampling and laboratory procedures can perturb their dispersion state or bias the distribution of other properties. In environmental contexts, an additional challenge is that many methods cannot detect low concentrations of nanoparticles that may still have an adverse effect. For some applications, nanoparticles may be characterized in complex matrices such as water, soil, food, polymers, inks, complex mixtures of organic liquids such as in cosmetics, or blood.

There are several overall categories of methods used to characterize nanoparticles. Microscopy methods generate images of individual nanoparticles to characterize their shape, size, and location. Electron microscopy and scanning probe microscopy are the dominant methods. Because nanoparticles have a size below the diffraction limit of visible light, conventional optical microscopy is not useful. Electron microscopes can be coupled to spectroscopic methods that can perform elemental analysis. Microscopy methods are destructive, and can be prone to undesirable artifacts from sample preparation, or from probe tip geometry in the case of scanning probe microscopy. Additionally, microscopy is based on single-particle measurements, meaning that large numbers of individual particles must be characterized to estimate their bulk properties.

Spectroscopy, which measures the particles' interaction with electromagnetic radiation as a function of wavelength, is useful for some classes of nanoparticles to characterize concentration, size, and shape. X-ray, ultraviolet–visible, infrared, and nuclear magnetic resonance spectroscopy can be used with nanoparticles. Light scattering methods using laser light, X-rays, or neutron scattering are used to determine particle size, with each method suitable for different size ranges and particle compositions. Some miscellaneous methods are electrophoresis for surface charge, the Brunauer–Emmett–Teller method for surface area, and X-ray diffraction for crystal structure, as well as mass spectrometry for particle mass, and particle counters for particle number. Chromatography, centrifugation, and filtration techniques can be used to separate nanoparticles by size or other physical properties before or during characterization.

Functionalization

Functionalization is the introduction of organic molecules or polymers on the surface of the nanoparticle. The surface coating of nanoparticles determines many of their physical and chemical properties, notably stability, solubility, and targeting. A coating that is multivalent or polymeric confers high stability. Functionalized nanomaterial-based catalysts can be used for catalysis of many known organic reactions.

An environmentally friendly and facile functionalization approach for synthesizing highly dispersed Graphene was developed. In this approach, graphenes were functionalized covalently with Gallic Acid (GA) in a one-pot free radical grafting method. This cost-effective method could be used for industrial mass production. The suspension of Gallic acid-functionalized graphene in water was test as a high performance heat transfer fluids for improved convective heat transfer in closed conduit flows.

Surface Coating for Biological Applications

For biological applications, the surface coating should be polar to give high aqueous solubility and prevent nanoparticle aggregation. In serum or on the cell surface, highly charged coatings promote non-specific binding, whereas polyethylene glycol linked to terminal hydroxyl or methoxy groups repel non-specific interactions. Nanoparticles can be linked to biological molecules that can act as address tags, to direct the nanoparticles to specific sites within the body, specific organelles within the cell, or to follow specifically the movement of individual protein or RNA molecules in living cells. Common address tags are monoclonal antibodies, aptamers, streptavidin or peptides. These targeting agents should ideally be covalently linked to the nanoparticle and should be present in a controlled number per nanoparticle. Multivalent nanoparticles, bearing multiple targeting groups, can cluster receptors, which can activate cellular signaling pathways, and give stronger anchoring. Monovalent nanoparticles, bearing a single binding site, avoid clustering and so are preferable for tracking the behavior of individual proteins.

Red blood cell coatings can help nanoparticles evade the immune system.

Health and Safety

Nanoparticles present possible dangers, both medically and environmentally. Most of these are due to the high surface to volume ratio, which can make the particles very reactive or catalytic. They are also able to pass through cell membranes in organisms, and their interactions with biological systems are relatively unknown. However, it is unlikely the particles would enter the cell nucleus, Golgi complex, endoplasmic reticulum or other internal cellular components due to the particle size and intercellular agglomeration. A recent study looking at the effects of ZnO nanoparticles on human immune cells has found varying levels of susceptibility to cytotoxicity. There are concerns that pharmaceutical companies, seeking regulatory approval for nano-reformulations of existing medicines, are relying on safety data produced during clinical studies of the earlier, pre-reformulation version of the medicine. This could result in regulatory bodies, such as the FDA, missing new side effects that are specific to the nano-reformulation. However considerable research has demonstrated that zinc nanoparticles are not absorbed into the bloodstream in vivo.

Concern has also been raised over the health effects of respirable nanoparticles from certain combustion processes. As of 2013 the U.S. Environmental Protection Agency was investigating the safety of the following nanoparticles:

- Carbon Nanotubes: Carbon materials have a wide range of uses, ranging from composites for use in vehicles and sports equipment to integrated circuits for electronic components. The interactions between nanomaterials such as carbon nanotubes and natural organic matter strongly influence both their aggregation and deposition, which strongly affects their transport, transformation, and exposure in aquatic environments. In past research, carbon nanotubes exhibited some toxicological impacts that will be evaluated in various environmental settings in current EPA chemical safety research. EPA research will provide data, models, test methods, and best practices to discover the acute health effects of carbon nanotubes and identify methods to predict them.
- Cerium oxide: Nanoscale cerium oxide is used in electronics, biomedical supplies, energy, and fuel additives. Many applications of engineered cerium oxide nanoparticles naturally disperse themselves into the environment, which increases the risk of exposure. There is ongoing exposure to new diesel emissions using fuel additives containing CeO_2 nanoparticles, and the environmental and public health impacts of this new technology are unknown. EPA's chemical safety research is assessing the environmental, ecological, and health implications of nanotechnology-enabled diesel fuel additives.
- Titanium dioxide: Nano titanium dioxide is currently used in many products. Depending on the type of particle, it may be found in sunscreens, cosmetics, and paints and coatings. It is also being investigated for use in removing contaminants from drinking water.

- Nano Silver: Nano silver is being incorporated into textiles, clothing, food packaging, and other materials to eliminate bacteria. EPA and the U.S. Consumer Product Safety Commission are studying certain products to see whether they transfer nano-size silver particles in real-world scenarios. EPA is researching this topic to better understand how much nano-silver children come in contact with in their environments.
- Iron: While nano-scale iron is being investigated for many uses, including "smart fluids" for uses such as optics polishing and as a better-absorbed iron nutrient supplement, one of its more prominent current uses is to remove contamination from groundwater. This use, supported by EPA research, is being piloted at a number of sites across the United States.

Applications

As the most prevalent morphology of nanomaterials used in consumer products, nanoparticles have potential and actual applications in all industries. Table below summarizes the most common nanoparticles used in various product types available on the global markets.

Table: Various nanoparticles which are commonly used in the consumer products by industrial sectors.

Industrial sectors	Nanoparticles
Agriculture	Silver, silicon dioxide, potassium, calcium, iron, zinc, phosphorus, boron, zinc oxide and molybdenum.
Automotive	Tungsten, disulfidesilicon dioxide, clay, titanium dioxide, diamond, copper, cobalt oxide, zinc oxide, boron nitride, zirconium dioxide, tungsten, γ-aluminium oxide, boron, palladium, platinum, cerium(IV) oxide, carnauba, aluminium oxide, silver, calcium carbonate and calcium sulfonate.
Construction	Titanium, dioxidesilicon dioxide, silver, clay, aluminium oxide, calcium carbonate calcium silicate hydrate, carbon, aluminium phosphate cerium(IV) oxide and calcium hydroxide.
Cosmetics	Silver, titanium dioxide, gold, carbon, zinc oxide, silicon dioxide, clay, sodium silicate, kojic acid and hydroxy acid.
Electronics	Silver, aluminum, silicon dioxide and palladium.
Environment	Silver, titanium dioxide, carbonmanganese oxide, clay, gold and selenium.
Food	Silver, clay, titanium dioxide, gold, zinc oxide, silicon dioxide, calcium, copper, zinc, platinum, manganese, palladium and carbon.
Home Appliance	Silver, zinc oxide, silicon dioxide, diamond and titanium dioxide.
Medicine	Silver, gold, hydroxyapatite, clay, titanium dioxide, silicon dioxide, zirconium dioxide, carbon, diamond, aluminium oxide and ytterbium trifluoride.
Petroleum	Tungsten, disulfidezinc oxide, silicon dioxide, diamond, clay, boron, boron nitride, silver, titanium dioxide, tungsten, γ-aluminium oxide, carbon, molybdenum disulfide and γ-aluminium oxide.

Printing	Toner, deposited by a printer onto paper or other substrate.
Renewable Energies	Titanium, palladium, tungsten disulfide, silicon dioxide, clay, graphite, zirconium(IV) oxide-yttria stabilized, carbon, gd-doped-cerium(IV) oxide, nickel cobalt oxide, nickel(II) oxide, rhodium, sm-doped-cerium(IV) oxide, barium strontium titanate and silver.
Sports and Fitness	Silver, titanium dioxide, gold, clay and carbon.
Textile	Silver, carbon, titanium dioxide, copper sulfide, clay, gold, polyethylene terephthalate and silicon dioxide.

Scientific research on nanoparticles is intense as they have many potential applications in medicine, physics, optics, and electronics. The U.S. National Nanotechnology Initiative offers government funding focused on nanoparticle research. The use of nanoparticles in laser dye-doped poly(methyl methacrylate) (PMMA) laser gain media was demonstrated in 2003 and it has been shown to improve conversion efficiencies and to decrease laser beam divergence. Researchers attribute the reduction in beam divergence to improved dn/dT characteristics of the organic-inorganic dye-doped nanocomposite. The optimum composition reported by these researchers is 30% w/w of SiO2 (~ 12 nm) in dye-doped PMMA. |Nanoparticles are being investigated as potential drug delivery system. Drugs, growth factors or other biomolecules can be conjugated to nano particles to aid targeted delivery. This nanoparticle-assisted delivery allows for spatial and temporal controls of the loaded drugs to achieve the most desirable biological outcome. Nanoparticles are also studied for possible applications as dietary supplements for delivery of biologically active substances, for example mineral elements.

SELF-ASSEMBLY OF NANOPARTICLES

The empirical definition of self-assembly is typically given as a phenomenon where the components of a system assemble themselves to form a larger functional unit. This spontaneous organization can be due to direct specific interaction, collective effects, and/or occur indirectly through their environment. The thermodynamics-based definition of self-assembly was introduced by Nicholas A. Kotov and describes self-assembly as a process where components of the system acquire non-random spatial distribution with respect to each other and the boundaries of the system. This definition allows one to account for mass and energy fluxes taking place in the self-assembly processes and is specifically applicable to the processes of spontaneous organization of nanoparticles with each other and with the boundaries of the system.

Due to the proliferation of nanoparticle synthesis techniques, the study and design of nanoparticle self-assembly has become widespread. The spatial arrangements of these self-assembled nanoparticles can be potentially used to build increasingly complex structures leading to a wide variety of materials that can be used for different purposes.

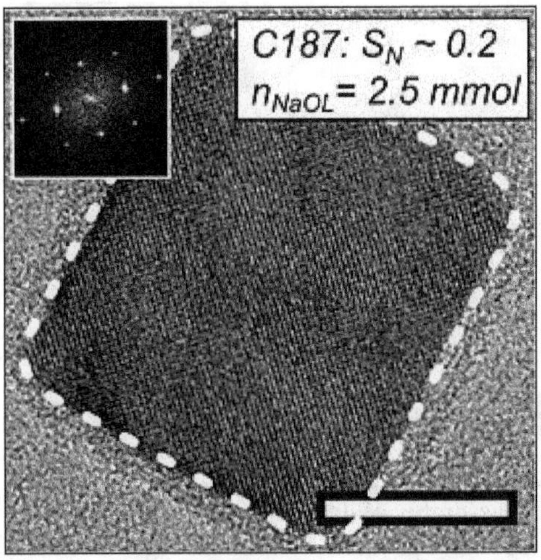

Transmission electron microscopy image of an iron oxide nanoparticle. Regularly arranged dots within the dashed border are columns of Fe atoms. Left inset is the corresponding electron diffraction pattern. Scale bar: 10 nm.

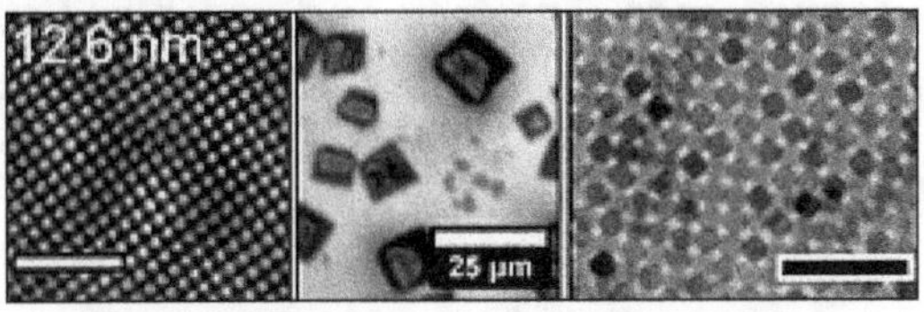

Iron oxide nanoparticles can be dispersed in an organic solvent (toluene). Upon its evaporation, they may self-assemble (left and right panels) into micron-sized mesocrystals (center) or multilayers (right). Each dot in the left image is a traditional "atomic" crystal shown in the image above. Scale bars: 100 nm (left), 25 μm (center), 50 nm (right).

At the molecular level, intermolecular force hold the spontaneous gathering of molecules into a well-defined and stable structure together. In chemical solutions, self-assembly is an outcome of random motion of molecules and the affinity of their binding sites for one another. In the area of nanotechnology, developing a simple, efficient method to organize molecules and molecular clusters into precise, pre-determined structure is crucial.

In 1959, physicist Richard Feynman gave a talk titled *There's Plenty of Room at the Bottom* to the American Physical Society. He imagined a world in which "we could arrange atoms one by one, just as we want them." This idea set the stage for the bottom-up synthesis approach in which constituent components interact to form higher-ordered structures in a controllable manner. The study of self-assembly of nanoparticles began with recognition that some properties of atoms and molecules enable them to arrange themselves into patterns. A variety of applications where the self-assembly of nanoparticles might be useful. For example, building sensors or computer chips. Nanoparticles have been observed by transmission electron microscopy (TEM), scanning electron microscopy (SEM), and X-ray diffraction to self-assemble in real-time.

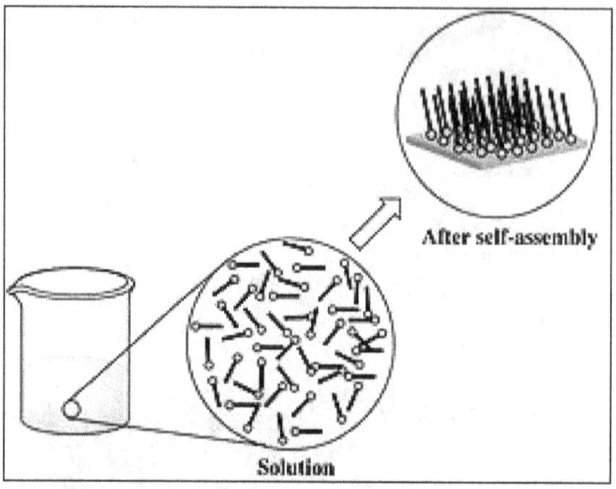

An example of self-assembly of nanoparticles in a solution. In this diagram, it can be seen that a disordered system formed an organized structure which can be due to specific interactions among the particles.

Self-assembly is defined as a process in which individual units of material associate with themselves spontaneously into a defined and organized structure or larger units with minimal external direction. Self-assembly is recognized as a highly useful technique to achieve outstanding qualities in both organic and inorganic nanostructures.

Importance

Self assembly of nanomaterials is currently considered broadly for nano-structuring and nano-fabrication because of its simplicity, versatility and spontaneity. Exploiting the properties of the nano assembly holds promise as a low-cost and high-yield technique for a wide range of scientific and technological applications and is a key research effort in nanotechnology, molecular robotics, and molecular computation. A summary of benefits of self-assembly in fabrication is listed below:

- Self-assembly is a scalable and parallel process which can involve large numbers of components in a short timeframe.
- Can result in structural dimensions across orders of magnitude, from nanoscale to macroscale.
- Is relatively inexpensive compared to the top-down assembly approach, which often consumes large amounts of finite resources.
- Natural processes that drive self-assembly tend to be highly reproducible.

Challenges

There exist several outstanding challenges in self-assembly, as it is a relatively new field in nanotechnology. Currently self-assembly is difficult to control on large scales, and to be widely applied we will need to ensure high degrees of reproducibility at these scales. The fundamental thermodynamic and kinetic mechanisms of self-assembly are poorly understood - the basic principles of atomistic and macroscale processes can be significantly different than those for nanostructures. Concepts related to thermal motion and capillary action influence equilibrium timescales and kinetic rates that are not well defined in self-assembling systems.

Top-down vs. Bottom-up Synthesis

A bottom up approach for nano-assembly is a primary research target for nano-fabrication because top down synthesis is expensive (requiring external work) and is not selective on very small lengthscales, but is currently the primary mode of industrial fabrication. Generally,the maximum resolution of the top-down products is much coarser than those of bottom-up; therefore, an accessible strategy to bridge "bottom-up" and "top-down," is realizable by the principles of self-assembly. By controlling local intermolecular forces to find the lowest-energy configuration, self-assembly can be guided by templates to generate similar structures to those currently fabricated by top-down approaches. This so-called bridging will enable fabrication of materials with the fine resolution of bottom-up methods and the larger range and arbitrary structure of top-down processes. Furthermore, in some cases components are too small for top-down synthesis, so self-assembly principles are required to realize these novel structures.

Classification

Nanostructures can be organized into groups based on their size, function, and structure; this organization is useful to define the potential of the field.

By Size

Among the more sophisticated and structurally complex nanostructures currently available are organic macromolecules, wherein their assembly relies on the placement of atoms into molecular or extended structures with atomic-level precision. It is now known that organic compounds can be conductors, semiconductors, and insulators, thus one of the main opportunities in nanomaterials science is to use organic synthesis and molecular design to make electronically useful structures. Structural motifs in these systems include colloids, small crystals, and aggregates on the order of 1-100 nm.

By Function

Nanostructured materials can also be classed according to their functions, for example nanoelectronics and information technology (IT). Lateral dimensions used in information storage are shrinking from the micro- to the nanoscale as fabrication technologies improve. Optical materials are important in the development of miniaturized information storage because light has many advantages for storage and transmission over electronic methods. Quantum dots - most commonly CdSe nanoparticles having diameters of tens of nm, and with protective surface coatings - are notable for their ability to fluoresce over a broad range of the visible spectrum, with the controlling parameter being size.

By Structure

Certain structural classes are especially relevant to nanoscience. As the dimensions of structures become smaller, their surface area-to-volume ratio increases. Much like molecules, nanostructures at small enough scales are essentially "all surface". The mechanical properties of materials are strongly influenced by these surface structures. Fracture strength and character, ductility, and various mechanical moduli all depend on the substructure of the materials over a range of scales. The opportunity to redevelop a science of materials that are nanostructured by design is largely open.

Thermodynamics

Self-assembly is an equilibrium process, i.e. the individual and assembled components exist in equilibrium. In addition, the flexibility and the lower free energy conformation is usually a result of a weaker intermolecular force between self-assembled moieties and is essentially enthalpic in nature.

The thermodynamics of the self-assembly process can be represented by a simple Gibbs free energy equation:

$$\Delta G_{SA} = \Delta H_{SA} - T\Delta S_{SA}$$

where if ΔG_{SA} is negative, self-assembly is a spontaneous process. ΔH_{SA} is the enthalpy change of the process and is largely determined by the potential energy/intermolecular forces between the assembling entities. ΔS_{SA} is the change in entropy associated with the formation of the ordered arrangement. In general, the organization is accompanied by a decrease in entropy and in order for the assembly to be spontaneous the enthalpy term must be negative and in excess of the entropy term. This equation shows that as the value of $T\Delta S_{SA}$ approaches the value of ΔH_{SA} and above a critical temperature, the self-assembly process will become progressively less likely to occur and spontaneous self-assembly will not happen.

The self-assembly is governed by the normal processes of nucleation and growth. Small assemblies are formed because of their increased lifetime as the attractive interactions between the components lower the Gibbs free energy. As the assembly grows, the Gibbs free energy continues to decrease until the assembly becomes stable enough to last for a long period of time. The necessity of the self-assembly to be an equilibrium process is defined by the organization of the structure which requires non-ideal arrangements to be formed before the lowest energy configuration is found.

Kinetics

The ultimate driving force in self-assembly is energy minimization and the corresponding evolution towards equilibrium, but kinetic effects can also play a very strong role. These kinetic effects, such as trapping in metastable states, slow coarsening kinetics, and pathway-dependent assembly, are often viewed as complications to be overcome in, for example, the formation of block copolymers.

Amphiphile self-assembly is an essential bottom-up approach of fabricating advanced functional materials. Self-assembled materials with desired structures are often obtained through thermodynamic control. Here, we demonstrate that the selection of kinetic pathways can lead to drastically different self-assembled structures, underlining the significance of kinetic control in self-assembly.

Defects

Self-assembled structure contain defects. Dislocations caused during the assembling of nanomaterials can majorly affect the final structure and in general defects are never completely avoidable. Current research on defects is focused on controlling defect density. In most cases, the thermodynamic driving force for self-assembly is provided by weak intermolecular interactions and is usually of the same order of magnitude as the entropy term. In order for a self-assembling system to reach the minimum free energy configuration, there has to be enough thermal energy to allow the mass transport of the self-assembling molecules. For defect formation, the free energy of single defect formation is given by:

$$\Delta G_{DF} = \Delta H_{DF} - T\Delta S_{DF}$$

The enthalpy term, ΔH_{DF} does not necessarily reflect the intermolecular forces between the molecules, it is the energy cost associated with disrupting the pattern and may be thought of as a region where optimum arrangement does not occur and the reduction of enthalpy associated with ideal self-assembly did not occur. An example of this can be seen in a system of hexagonally packed cylinders where defect regions of lamellar structure exist.

If $\ddot{A}G_{DF}$ is negative, there will be a finite number of defects in the system and the concentration will be given by:

$$\frac{N}{N_0} = \exp(\frac{-\Delta E_{act}}{RT})$$

N is the number of defects in a matrix of N_0 self-assembled particles or features and ΔE_{act} is the activation energy of defect formation. The activation energy, ΔE_{act} , should not be confused with $\ddot{A}H_{DF}$. The activation energy represents the energy difference between the initial ideally arranges state and a transition state towards the defective structure. At low defect concentrations, defect formation is entropy driven until a critical concentration of defects allows the activation energy term to compensate for entropy. There is usually an equilibrium defect density indicated at the minimum free energy. The activation energy for defect formation increases this equilibrium defect density.

Particle Interaction

Intermolecular forces govern the particle interaction in self-assembled systems. The forces tend to be intermolecular in type rather than ionic or covalent because ionic or covalent bonds will "lock" the assembly into non-equilibrium structures. The types intermolecular forces seen in self-assembly processes are van der Waals, hydrogen bonds, and weak polar forces, just to name a few. In self-assembly, regular structural arrangements are frequently observed, therefore there must be a balance of attractive and repulsive between molecules otherwise an equilibrium distance will not exist between the particles. The repulsive forces can be electron cloud-electron cloud overlap or electrostatic repulsion.

Processing

The processes by which nanoparticles self-assemble are widespread and important. Understanding why and how self-assembly occurs is key in reproducing and optimizing results. Typically, nanoparticles will self-assemble for one or both of two reasons: molecular interactions and external direction.

Self-assembly by Molecular Interactions

Nanoparticles have the ability to assemble chemically through covalent or noncovalent interactions with their capping ligand. The terminal functional group(s) on the particle are known as capping ligands. As these ligands tend to be complex and sophisticated, self-assembly can provide a simpler pathway for nanoparticle organization by synthesizing efficient functional groups. For instance, DNA oligomers have been a key ligand for nanoparticle building blocks to be self-assembling via sequence-based specific organization. However, to deliver precise and scalable (programmable) assembly for a desired structure, a careful positioning of ligand molecules onto the nanoparticle counterpart should be required at the building block (precursor) level, such as direction, geometry, morphology, affinity, etc. The successful design of ligand-building block units can play an essential role in manufacturing a wide-range of new nano systems, such as nanosensor systems, nanomachines/nanobots, nanocomputers, and many more uncharted systems.

Intermolecular Forces

Nanoparticles can self-assemble as a result of their intermolecular forces. As systems look to minimize their free energy, self-assembly is one option for the system to achieve its lowest free energy thermodynamically. Nanoparticles can be programmed to self-assemble by changing the functionality of their side groups, taking advantage of weak and specific intermolecular forces to spontaneously order the particles. These direct interparticle interactions can be typical intermolecular forces such as hydrogen bonding or Van der Waals forces, but can also be internal characteristics, such as hydrophobicity or hydrophilicity. For example, lipophilic nanoparticles have the tendency to self-assemble and form crystals as solvents are evaporated. While these aggregations are based on intermolecular forces, external factors such as temperature and pH also play a role in spontaneous self-assembly.

Hamaker Interaction

As nanoparticle interactions take place on a nanoscale, the particle interactions must be scaled similarly. Hamaker interactions take into account the polarization characteristics of a large number of nearby particles and the effects they have on each other. Hamaker interactions sum all of the forces between all particles and the solvent(s) involved in the system. While Hamaker theory generally describes a macroscopic system, the vast number of nanoparticles in a self-assembling system allows the term to be applicable. Hamaker constants for nanoparticles are calculated using Lifshitz theory, and can often be found in literature.

Table: Hamaker constants for nanoparticles in water.

Material	A_{131}
Fe_3O_4	22
γ-Fe_2O_3	26
α-Fe_2O_3	29
Ag	33
Au	45
All values reported in zJ	

Externally Directed Self-assembly

The natural ability of nanoparticles to self-assemble can be replicated in systems that do not intrinsically self-assemble. Directed self-assembly (DSA) attempts to mimic the chemical properties of self-assembling systems, while simultaneously controlling the thermodynamic system to maximize self-assembly.

Electric and Magnetic Fields

External fields are the most common directors of self-assembly. Electric and magnetic fields allow induced interactions to align the particles. The fields take advantage of the polarizability of the nanoparticle and its functional groups. When these field-induced interactions overcome random Brownian motion, particles join to form chains and then assemble. At more modest field strengths, ordered crystal structures are established due to the induced dipole interactions. Electric and magnetic field direction requires a constant balance between thermal energy and interaction energies.

Flow Fields

Common ways of incorporating nanoparticle self-assembly with a flow include Langmuir-Blodgett, dip coating, flow coating and spin coating.

Macroscopic Viscous Flow

Macroscopic viscous flow fields can direct self-assembly of a random solution of particles into ordered crystals. However, the assembled particles tend to disassemble when the flow is stopped or removed. Shear flows are useful for jammed suspensions or random close packing. As these systems begin in nonequilibrium, flow fields are useful in that they help the system relax towards ordered equilibrium. Flow fields are also useful when dealing with complex matrices that themselves have rheological behavior. Flow can induce anisotropic viseoelastic stresses, which helps to overcome the matrix and cause self-assembly.

Large Amplitude Oscillatory Shear (LAOS)

Large amplitude oscillatory shear (LAOS) is most effective for particles that are 100 nm-1 μm in size. Hard and soft shears can order in steady shear. However, this assembly strongly relies on particle volume fraction, particle interaction potentials, polydisterity, and shear rate and strain. The large amount of directing factors can cause complications in directing self-assembly by LAOS. Diblock copolymer micells have been studied in regard to structuring nanoparticles in bulk.

Combination of Fields

The most effective self-assembly director is a combination of fields. If the fields and conditions are optimized, self-assembly can be permanent and complete. When a field combination is used with nanoparticles that are tailored to be intrinsically responsive, the most complete assembly is observed. Combinations of fields allow the benefits of self-assembly, such as scalability and simplicity, to be maintained while being able to control orientation and structure formation. Field combinations possess the greatest potential for future directed self-assembly work.

Interfaces

Solid Interfaces

Nano-particles can self-assemble on solid surfaces after applying external forces (like magnetic, electric, and flow). Templates made of microstructures like carbon nanotubes or block polymers can also be used to assist in self-assembly; they cause directed self-assembly (DSA) in which active sites are embedded to selectively induce nanoparticle deposition. Such templates are considered as any object onto which different particles can be arranged into a structure with a morphology similar to that of the template. Carbon nanotubes (microstructures), single molecules, or block copolymers are common templates. Nanoparticles are often shown to self-assemble within distances of nanometers and micrometers, but block copolymer templates can be used to form well-defined self-assemblies over macroscopic distances. By incorporating active sites to the surfaces of nanotubes and polymers, the functionalization of these templates can be transformed to favor self-assembly of specified nanoparticles.

Liquid Interfaces

Pickering and Ramsden explained the idea of pickering emulsions when experimenting with paraffin-water emulsions with solid particles like iron oxide and silicon dioxide. They observed that the micron-sized colloids generated a resistant film at the interface between the two immiscible phases, inhibiting the coalescence of the emulsion drops. These Pickering emulsions, as shown in the figure below, are formed from the self-assembly of colloidal particles in two-part liquid systems, such as oil-water systems. The desorption energy, which is directly related to the stability of emulsions depends on the particle size, particle-particle interaction and, of course, particle-water and particle-oil interactions.

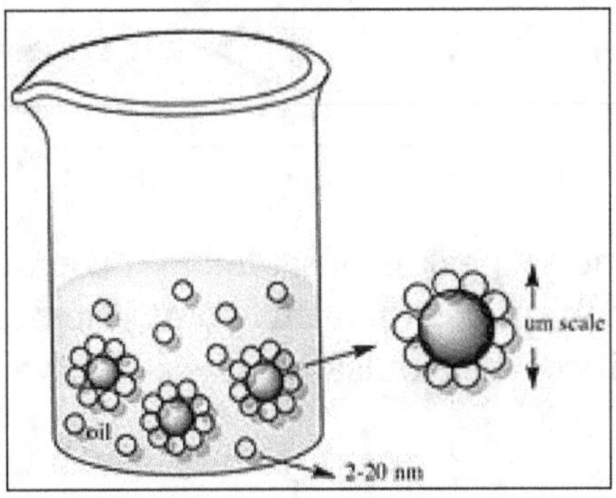

Self-assembly of solid nanoparticles at oil-water interface.

A decrease in total free energy was observed to be a result from the assembly of nanoparticles at an oil/water (O/W) interface. This is shown in the following equation in which a particle with radius r at an interface between oil (O) and water (W) results in a decrease of the initial interfacial energy E_0 to E_1; this difference in energy is ΔE_1.

$$E_0 - E_1 = -\frac{\pi r^2}{\gamma_{O/W}}[\gamma_{O/W} - (\gamma_{P/W} - \gamma_{P/O})]^2$$

At a fixed $\gamma_{P/O}$, $\gamma_{P/W}$, and $\gamma_{O/W}$, the equation shows that the stability of a particle assembly is determined by the radius square. When moving to the interface, particles reduce the unfavorable contact between the immiscible fluids and decrease the interfacial energy. The decrease in total free energy for microscopic particles is much larger than that of thermal energy; this results in an effective confinement of large colloids to the interface. They are irreversibly bound to the interface. Nanoscopic particles are restricted to the interface by an energy reduction comparable to thermal energy. Thus, nanoparticles are easily displaced from the interface. A constant particle exchange then occurs at the interface; the rate of this exchange depends on particle size. The thermally activated escape of small particles occurs more often than larger particles. For the equilibrium state of assembly, the total gain in free energy is smaller for smaller particles. Thus, large nanoparticles assemblies are more stable. This size dependence allows nanoparticles to self-assemble at the interface to attain its equilibrium structure. Micrometer-size colloids, on the other hand, may be confined in a non-equilibrium state.

The interfacial tension and the wettability of a particle surface affect the desorption energy. The contact angle θ between the solid and the oil/water interface determines its wettability. As shown in the figure below, a contact angle θ greater than 90° favors a water-in-oil emersion while a contact angle θ less than 90° favors oil-in-water emulsion. These contact angles affect the stability of the emulsion.

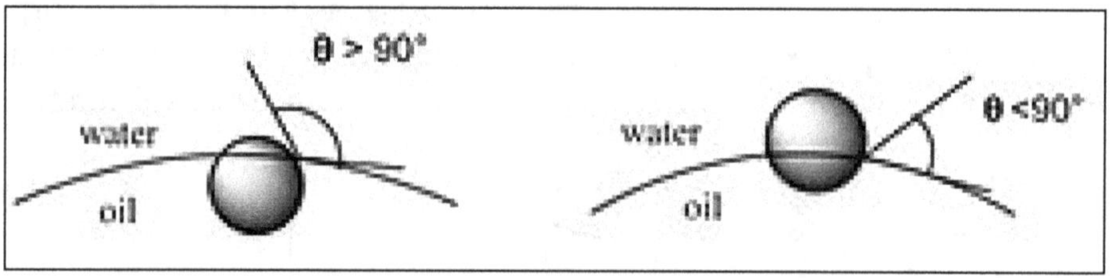

Effects of contact angle on wettability.

A maximum desorption energy peak is observed at a contact angle of 90°. When the contact angle is greater than or less than this point, the desorption energy gradually decreases; thus, the stability of the emulsion decreases as well.

Applications

Material that consists with nano-particles is called "nanostructured material". The phase of "nanostructured material" implies two important ideas:

- That at least some of the heterogeneity in materials is determined by the size range of nanostructures (about 1–100 nm).
- These nanostructures might be synthesized and distributed (or organized), at least in part. The study of self-assembly nanoparticles is important to understand the interaction between single particle in terms of applying them into different applications. As the field is growing rapidly, several applications in nano electronics and bioelectronics have yet to be achieved and applied in the real world.

Electronics

Self-assembly of nanoscale structures from functional nanoparticles has provided a powerful path to developing small and powerful electronic components. The difficulty of applied nanostructure material is nanoscale objects have always been difficult to manipulate because they cannot be characterized by molecular techniques and they are too small to observe optically. 2D self-assembly monodisperse particle colloids has a strong potential in dense magnetic storage media. Each colloid particle has the ability to store information as known as binary number 0 and 1 after applying it to a strong magnetic field. In the meantime, it requires a nanoscale sensor or detector in order to selectively choose the colloid particle.

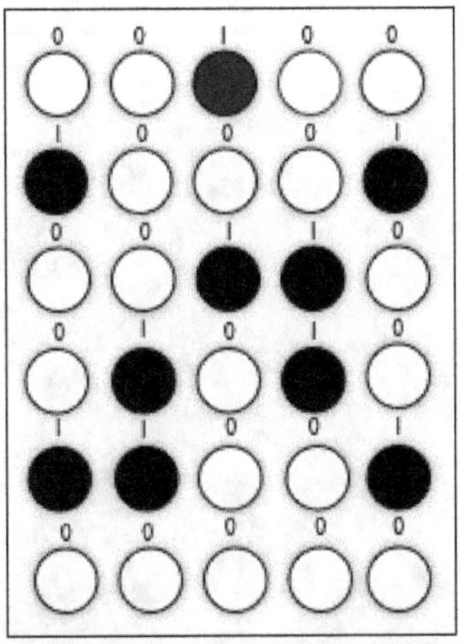

Model of multidimensional array of nano-particles. A particle could have two spins, spin up or down. Based on the spin directions, nano-particles will be able to store 0 and 1. Therefore, nanostructural material has a great potential for future use in electronic devices.

Biological Applications

Drug Delivery

Block copolymers are a well-studied and versatile class of self-assembling materials characterized by chemically distinct polymer blocks that are covalently bonded. This molecular architecture of the covalent bond enhancement is what causes block copolymers to spontaneously form nanoscale patterns. These copolymers are a well-studied and versatile class of self-assembling materials characterized by chemically distinct polymer blocks that are covalently bonded. This molecular architecture of the covalent bond enhancement, is what causes block copolymers to spontaneously form nanoscale patterns. These copolymers offer the ability to self-assemble into uniform, nanosized micelles and accumulate in tumors via the enhanced permeability and retention effect. Polymer composition can be chosen to control the micelle size and compatibility with the drug of choice. The challenges of this application are the difficulty of reproducing or controlling the size of self-assembly nano micelle, preparing predictable size-distribution, and the stability of the micelle with high drug load content.

Magnetic Drug Delivery

Magnetic nanochains are a class of new magnetoresponsive and superparamagnetic nanostructures with highly anisotropic shapes (chain-like) which can be manipulated using magnetic field and magnetic field gradient. The magnetic nanochains possess attractive properties which are significant added value for many potential uses including magneto-mechanical actuation-associated nanomedicines in low and super-low frequency alternating magnetic field and magnetic drug delivery.

NANOPARTICLE DEPOSITION

Nanoparticle coating of polystyrene nanoparticles on quartz prepared with the Langmuir-Blodgett method.

Nanoparticle deposition refers to the process of attaching nanoparticles to solid surfaces called substrates to create coatings of nanoparticles. The coatings can have a monolayer or a multilayer and organized or unorganized structure based on the coating method used. Nanoparticles are typically difficult to deposit due to their physical properties.

Challenges

Nanoparticles can be made from different materials such as metals, ceramics and polymers. The stability of the nanoparticles can be an issue as nanoparticles have a tendency to lower their very high surface energy, which originates from their high surface-to-bulk ratio. Bare nanoparticles tend to stabilize themselves either by sorption of molecules from the surroundings or by lowering the surface area through coagulation and agglomeration. Usually the formation of these aggregates is unwanted. The tendency of a nanoparticle to coagulate can be controlled by modifying the surface layer. In a liquid medium, suitable ligand molecules are commonly attached to the nanoparticle surface, as they provide solubility in suitable solvents and prevent coagulation.

Deposition Methods

There are multiple different coating methods available to deposit nanoparticles. The methods differ by their ability to control particle packing density and layer thickness, ability to use different particles and the complexity of the method and the instrumentation needed.

Langmuir-Blodgett

In the Langmuir-Blodgett method, the nanoparticles are injected at air-water interphase in a special Langmuir-Blodgett Trough. The floating particles are compressed closer to each other with motorized barriers which allow to control the packing density of the particles. After compressing the particles to the desired packing density, they are transferred on a solid substrate using vertical (Langmuir-Blodgett) or horizontal (Langmuir-Schaefer) dipping to create a monolayer coating. Controlled multilayer coatings can be made repeating the dipping procedure multiple times.

The benefits of the Langmuir-Blodgett method include a firm control over the packing density and the layer thickness achieved that have been shown to be better than with other methods, the ability to use different shapes and materials of substrates and particles and the possibility to characterize the particle layer during deposition for example a Brewster Angle Microscope. As a disadvantage, a successful Langmuir-Blodgett deposition requires optimization of multiple measurement parameters such as dipping speed, temperature and dipping packing density.

Dip Coating and Spin Coating

The spin and dip coating methods are simple methods for nanoparticle deposition. They are useful tools especially in creating self-assembled layers and films where the packing density isn't critical. Accurate and vibration-free sample withdrawal speeds can be used to have control over the film thickness. Creating high density monolayers is typically very difficult since the methods are lacking the packing density control. Also, the volume of nanoparticle suspension required for both spin coating and dip coating is rather big which may be an issue when using expensive nanoparticle materials.

Other Methods

Other possible deposition methods include methods utilizing particle self-assembly by solvent evaporation, doctor blade, chemical vapor deposition and transfer printing. Some of these methods like solvent evaporation are extremely simple but produce low-quality films. Other methods such as the chemical vapor deposition are effective for certain types of particles and substrates but are limited in particle types that can be used and require heavier instrumentation investments. Also hybrid methods such as combining self-assembly to Langmuir-Blodgett have been used.

Nanoparticle Coating Applications

Coatings and thin films made from nanoparticles are being used in various applications including displays, sensors, medical devices, energy storages and energy harvesting. Examples include:

- Using graphene oxide for applications in electronics.
- Using nanoparticles of metal oxides, carbon nanotubes and quantum dots in photovoltaics, displays and sensors.
- Using polymers and nanocomposites in nanolithographic patterning.

NANOPARTICLE DRUG DELIVERY

Nanoparticle drug delivery systems are engineered technologies that use nanoparticles for the targeted delivery and controlled release of therapeutic agents. The modern form of a drug delivery system should minimize side-effects and reduce both dosage and dosage frequency. Recently, nanoparticles have aroused attention due to their potential application for effective drug delivery. Nanomaterials exhibit different chemical and physical properties or biological effects compared to larger-scale counterparts that can be beneficial for drug delivery systems. Some important advantages of nanoparticles is their high surface-area-to-volume ratio, chemical and geometric tunability, and their ability to interact with biomolecules to facilitate uptake across the cell membrane. The large surface area also has a large affinity for drugs and small molecules, like ligands or antibodies, for targeting and controlled release purposes. Nanoparticles refer to a large family of materials both organic and inorganic. Each material has uniquely tunable properties and thus can be selectively designed for specific applications. Despite the many advantages of nanoparticles, there are also many challenges, including but not exclusive to: nanotoxicity, biodistribution and accumulation, and the clearance of nanoparticles by human body. The National Institute of Biomedical Imaging and Bioengineering has issued the following prospects for future research in nanoparticle drug delivery systems:

- Crossing the blood-brain barrier (BBB) in brain diseases and disorders.
- Enhancing targeted intracellular delivery to ensure the treatments reach the correct structures inside cells.
- Combining diagnosis and treatment.

However, the development of a new drug system is relatively time-consuming; it takes approximately seven years to complete fundamental research and development before advancing to preclinical animal studies.

Characterization

Nanoparticle drug delivery focuses on maximizing drug efficacy and minimizing cytotoxicity. Fine-tuning nanoparticle properties for effective drug delivery involves addressing the following factors. The surface-area-to-volume ratio of nanoparticles can be altered to allow for more ligand binding to the surface. Increasing ligand binding efficiency can decrease dosage and minimize nanoparticle toxicity. Minimizing dosage or dosage frequency also lowers the mass of nanoparticle per mass of drug, thus achieving greater efficiency. Surface functionalization of nanoparticles is another important design aspect and is often accomplished by bioconjugation or passive adsorption of molecules onto the nanoparticle surface. By functionalizing nanoparticle surfaces with ligands that enhance drug binding, suppress immune response, or provide targeting/controlled release capabilities, both a greater efficacy and lower toxicity are achieved. Efficacy is increased as more drug is delivered to the target site, and toxic side effects are lowered by minimizing the total level of drug in the body. The composition of the nanoparticle can be chosen according to the target environment or desired effect. For example, liposome-based nanoparticles can be biologically degraded after delivery, thus minimizing the risk of accumulation and toxicity after the therapeutic cargo has been released. Metal nanoparticles, such as gold nanoparticles, have optical qualities(also described in nanomaterials) that allow for less invasive imaging techniques. Furthermore, the photothermal response of nanoparticles to optical stimulation can be directly utilized for tumor therapy.

Platforms

Current nanoparticle drug delivery systems can be cataloged based on their platform composition into several groups: polymeric nanoparticles, inorganic nanoparticles, viral nanoparticles, lipid-based nanoparticles, and nanoparticle albumin-bound (nab) technology. Each family has its unique characteristics.

Polymeric Nanoparticles

Polymeric nanoparticles are synthetic polymers with a size ranging from 10 to 100 nm. Common synthetic polymeric nanoparticles include polyacrylamide, polyacrylate, and chitosan. Drug molecules can be incorporated either during or after polymerization. Depending on the polymerization chemistry, the drug can be covalently bonded, encapsulated in a hydrophobic core, or conjugated electrostatically. Common synthetic strategies for polymeric nanoparticles include microfluidic approaches, electrodropping, high pressure homogenization, and emulsion-based interfacial polymerization. Polymer biodegradability is an important aspect to consider when choosing the appropriate nanoparticle chemistry. Nanocarriers composed of biodegradable polymers undergo hydrolysis in the body, producing biocompatible small molecules such as lactic acid and glycolic acid. Polymeric nanoparticles can be created via self-assembly or other methods such as particle replication in nonwetting templates (PRINT) which allows customization of composition, size, and shape of the nanoparticle using tiny molds.

Dendrimers

Dendrimers are unique hyper-branched synthetic polymers with monodispersed size, well-defined structure, and a highly functionalized terminal surface. They are typically composed of synthetic or natural amino acid, nucleic acids, and carbohydrates. Therapeutics can be loaded with relative ease onto the interior of the dendrimers or the terminal surface of the branches via electrostatic interaction, hydrophobic interactions, hydrogen bonds, chemical linkages, or covalent conjugation. Drug-dendrimer conjugation can elongate the half-life of drugs. Currently, dendrimer use in biological systems is limited due to dendrimer toxicity and limitations in their synthesis methods. Dendrimers are also confined within a narrow size range (<15 nm) and current synthesis methods are subject to low yield. The surface groups will reach the de Gennes dense packing limit at high generation level, which seals the interior from the bulk solution – this can be useful for encapsulation of hydrophobic, poorly soluble drug molecules. The seal can be tuned by intramolecular interactions between adjacent surface groups, which can be varied by the condition of the solution, such as pH, polarity, and temperature, a property which can be utilized to tailor encapsulation and controlled release properties.

Platinum nanoparticles.

Inorganic Nanoparticles

Inorganic nanoparticles have emerged as highly valuable functional building blocks for drug delivery systems due to their well-defined and highly tunable properties, such as size, shape, and surface functionalization. They have been largely adopted to biological and medical applications ranging from imaging and diagnoses to drug delivery. Drug delivery effectiveness also highly depends on the nanoparticle characteristics, which need to be thoroughly studied. Inorganic nanoparticles are typically metal nanoparticles. They are usually composed of inert metals such as gold and titanium that form nanoshells, although iron oxide nanoparticles have also become an option. Metal nanoparticles have special properties that can be utilized for smart drug delivery or incorporated into additional therapeutic effects. Because this platform is not composed of biodegradable components, there are concerns regarding toxicity and hypersensitivity. Challenges may arise from the interactions of these materials with biosystems, and with rapid clearance by human body. Despite the fact that these nanoparticles are composed of biocompatible and inert substances, a considerable amount of the particles may remain in the body after treatment, and the buildup of metal particles in the body after several applications may result in toxicity.

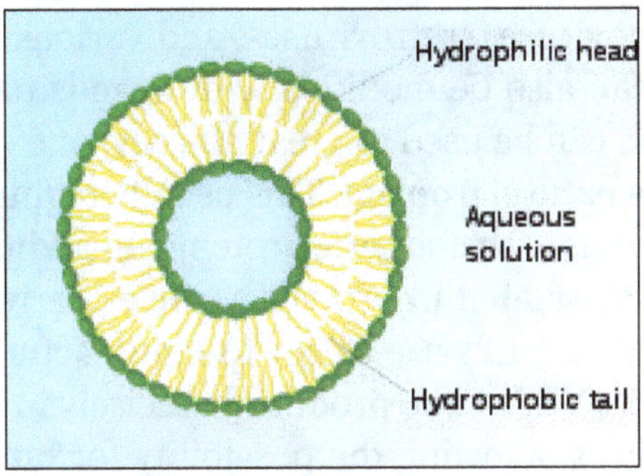

Basic liposome structure.

Liposome Delivery

Liposomes are spherical vesicles composed of synthetic or natural phospholipids that self-assemble in aqueous solution in sizes ranging from tens of nanometers to micrometers. The resulting vesicle, which has an aqueous core surrounded by a hydrophobic membrane, can be loaded with a wide variety of hydrophobic or hydrophilic molecules for therapeutic purposes. Liposomes are typically synthesized with naturally occurring phospholipids, mainly phosphatidylcholine. Cholesterol is often included in the formulation to adjust the rigidity of the membrane and to increase stability. The molecular cargo is loaded through liposome formation in aqueous solution, solvent exchange mechanisms, or pH gradients methods. Various molecules can also be chemically conjugated to the surface of the liposome to alter recognition properties. One typical modification is conjugating polyethyleneglycol (PEG) to the vesicle surface. The hydrophilic polymer prevents recognition by macrophages and decreases clearance. The size, surface charge, and bilayer fluidity also alter liposome delivery kinetics. Liposomes diffuse from the bloodstream into the interstitial space near the target site. As the cell membrane itself is composed of phospholipids, liposomes can directly fuse with the membrane and release the cargo into the cytosol, or may enter the cell through phagocytosis or other active transport pathways. Liposomal delivery has various advantages. Liposomes increase the solubility, stability, and uptake of drug molecules. Peptides, polymers, and other molecules can be conjugated to the surface of a liposome for targeted delivery. Conjugating various ligands can facilitate binding to target cells based on the receptor-ligand interaction. Altering vesicle size and surface chemistry can also be tuned to increase circulation time. Various FDA-approved liposomal drugs are in clinical use in the US. The anthracycline drug doxorubicin is delivered with phospholipid-cholesterol liposomes to treat AIDS-related Kaposi sarcoma and multiple myeloma with high efficacy and low toxicity. Many others are undergoing clinical trials, and liposomal drug delivery remains an active field of research today, with potential applications including nucleic acid therapy, brain targeting, and tumor therapy.

Viral Vectors, Viral-like Particles and Biological Nanocarriers

Viruses can be used to deliver genes for genetic engineering or gene therapy. Commonly used viruses include adenoviruses, retroviruses, and various bacteriophages. The surface of the viral particle can also be modified with ligands to increase targeting capabilities. While viral vectors can be used to great efficacy, one concern is that may cause off-target effects due to its natural tropism. This usually requires replacing the proteins causing virus-cell interactions with chimeric proteins. In addition to using viruses, drug molecules can also be encapsulated in protein particles derived from the viral capsid, or virus-like particles (VLPs). VLPs are easier to manufacture than viruses, and their structural uniformity allows VLPs to be produced precisely in large amounts. VLPs also have easy-to-modify surfaces, allowing the possibility for targeted delivery. There are various methods of packaging the molecule into the capsid; most take advantage of the capsid's ability to self-assemble. One strategy is to alter the pH gradient outside the capsid to create pores on the capsid surface and trap the desired molecule. Other methods use aggregators such as leucine zippers or polymer-DNA amphiphiles to induce capsid formation and capture drug molecules. It is also possible to chemically conjugate of drugs directly onto the reactive sites on the capsid surface, often involving the formation of amide bonds. After being introduced to the organism, VLPs often have broad tissue distribution, rapid clearance, and are generally non-toxic. It may, however, like viruses, invoke an immune response, so immune-masking agents may be necessary.

Nanoparticle Albumin-bound Technology

Nanoparticle albumin-bound technology utilizes the protein albumin as a carrier for hydrophobic chemotherapy drugs through noncovalent binding. Because albumin is already a natural carrier of hydrophobic particles and is able to transcytose molecules bound to itself, albumin composed nanoparticles have become an effective strategy for the treatment of many diseases in clinical research.

Delivery and Release Mechanisms

An ideal drug delivery system should have effective targeting and controlled release. The two main targeting strategies are passive targeting and active targeting. Passive targeting depends on the fact that tumors have abnormally structured blood vessels that favor accumulation of relatively large macromolecules and nanoparticles. This so-called enhanced permeability and retention effect (EPR) allows the drug-carrier be transported specifically to the tumor cells. Active targeting is, as the name suggests, much more specific and is achieved by taking advantage of receptor-ligand interactions at the surface of the cell membrane. Controlled drug release systems can be achieved through several methods. Rate-programmed drug delivery systems are tuned to the diffusivity of active agents across the membrane. Another delivery-release mechanism is activation-modulated drug delivery, where the release is triggered by environmental stimuli. The stimuli can be external, such as the introduction of a chemical activators or activation by light or electromagnetic fields, or biological - such as pH, temperature, and osmotic pressure which can vary widely throughout the body.

Polymeric Nanoparticles

For polymeric nanoparticles, the induction of stimuli-responsiveness has usually relied heavily upon well-known polymers that possess an inherent stimuli-responsiveness. Certain polymers that can undergo reversible phase transitions due to changes in temperature or pH have aroused interest. Arguably the most utilized polymer for activation-modulated delivery is the thermo-responsive polymer poly(N-isopropylacrylamide). It is readily soluble in water at room temperature but precipitates reversibly from when the temperature is raised above its lower critical solution temperature (LCST), changing from an extended chain conformation to a collapsed chain. This feature presents a way to change the hydrophilicity of a polymer via temperature. Efforts also focus on dual stimuli-responsive drug delivery systems, which can be harnessed to control the release of the encapsulated drug. For example, the triblock copolymer of poly(ethylene glycol)-b-poly(3-aminopropyl-methacrylamide)-b-poly(N-isopropylacrylamide) (PEG-b-PAPMA-b-PNIPAm) can self-assemble to form micelles, possessing a core–shell–corona architecture above the lower critical solution temperature. It is also pH responsive. Therefore, drug release can be tuned by changing either temperature or pH conditions.

Inorganic Nanoparticles

Drug delivery strategies of inorganic nanoparticles are dependent on material properties. The active targeting of inorganic nanoparticle drug carriers is often achieved by surface functionalization with specific ligands of nanoparticles. For example, the inorganic multifunctional nanovehicle (5-FU/Fe3O4/αZrP@CHI-FA-R6G) is able to accomplish tumor optical imaging and therapy simultaneously. It can be directed to the location of cancer cells with sustained release behavior. Studies have also been done on gold nanoparticle responses to local near-infrared (NIR) light as a stimuli for drug release. In one study, gold nanoparticles functionalized with double-stranded DNA encapsulated with drug molecules, were irradiated with NIR light. The particles generated heat and denatured the double-stranded DNA, which triggered the release of drugs at the target site. Studies also suggest that a porous structure is beneficial to attain a sustained or pulsatile release. Porous inorganic materials demonstrate high mechanical and chemical stability within a range of physiological conditions. The well-defined surface properties, such as high pore volume, narrow pore diameter distribution, and high surface area allow the entrapment of drugs, proteins and other biogenic molecules with predictable and reproducible release patterns.

Toxicity

Some of the same properties that make nanoparticles efficient drug carriers also contribute to their toxicity. For example, gold nanoparticles are known to interact with proteins through surface adsorption, which can be utilized for cargo loading and immune shielding. However, this protein-adsorption property can also disrupt normal protein function that is essential for homeostasis, especially when the protein contains exposed sulfur groups. The photothermal effect, which can be induced to kill tumor cells, may also create reactive oxygen species that impose oxidative stress on surrounding healthy cells. Gold nanoparticles of sizes below 4-5 nm fit in DNA grooves which can interfere with transcription, gene regulation, replication, and other processes that rely on DNA-protein binding. Lack of biodegradability for some nanoparticle chemistries can lead to accumulation in certain tissues, thus interfering with a wide range of biological processes. Currently, there is no regulatory framework in the United States for testing nanoparticles for their general impact on health and on the environment.

SILICON NANOTUBES

Silicon nanotubes are nanoparticles which create a tube-like structure from silicon atoms. As with silicon nanowires, they are technologically important due to their unusual physical properties, which differ fundamentally to those of bulk silicon. First reports on silicon nanotubes appeared around the year 2000.

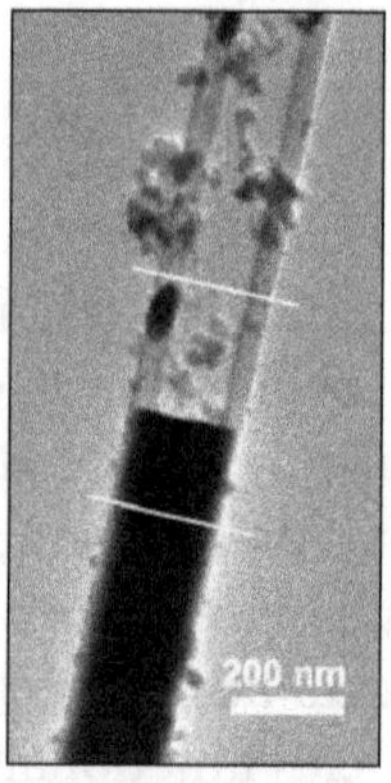

Si nanotube (top) created by partial etching of Si-covered ZnO nanowire (bottom).

Synthesis

One method to prepare silicon nanotubes is using a reactor employing an electric arc without the use of any catalyst. To ensure purity, the reactor is evacuated and filled with the nonreactive noble gas argon. The actual formation of the nanotubes relies on the process of chemical vapor deposition.

A more common laboratory-scale method involves the use of germanium, carbon or zinc oxide nanowires as a template. Silicon, coming typically from either silane or silicon tetrachloride gas, is then deposited onto the nanowires, and the core is dissolved leaving behind a silicon tube. The growth of template nanowires, silicon deposition and nanowire etching, and consequently the geometry of resulting Si nanotubes, can be accurately controlled in the second method; however, the smallest inner diameter is limited by tens of nanometers.

Si nanotubes created by etching of Si-covered ZnO nanowires.

The conventional vapor-liquid-solid (VLS) and solid-liquid-solid (SLS) mechanisms are favorite techniques to grow one-dimensional silicon nanostructures. However, they usually incorporate only one type of metal as catalyst and therefore can not be used for growing tubular (hollow) silicon nanostructures. In a recent attempt, a nickel-gold bilayer catalyst layer has been used to take the advantage of the uneven growth rate of constituent metal catalysts. Using these modified VLS and SLS techniques, multiwall silicon nanotubes with a sidewall thickness of few nanometers have been grown.

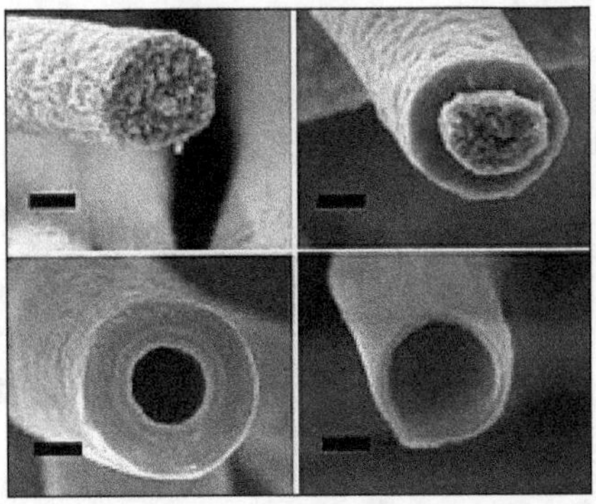

Si nanotubes produced using a carbon template. Clockwise: carbon fiber; carbon fiber coated with silicon; silicon oxide tube remaining after removing the carbon core; covering the silicon oxide with poly-crystalline silicon. Scale bars 200 nm.

Applications

As the result of Ballistic conductivity, silicon nanotubes and nanowires have been considered for use in electronics, e.g. in thermoelectric generators. Since the structure can accommodate molecules of hydrogen so it might resemble coal without the CO_2, it appears that silicon nanomaterials may behave like a metal fuel. A silicon nanotube charged with hydrogen delivers energy and in the process leaves residual water, ethanol, silicon and sand. However, as hydrogen production requires considerable energy, this is only a proposed method of storing energy, not producing it.

Silicon nanotubes and silicon nanowires can be used in lithium-ion batteries. Conventional Li-ion batteries use graphitic carbon as the anode, but replacing this with silicon nanotubes experimentally increases the specific (by mass) anode capacity by a factor of 10 (though the overall capacity improvement is lower due to the far lower specific cathode capacities).

Another emerging application of silicon nanotube is light emission. Since silicon is an indirect band gap semiconductor, the quantum yield of radiative recombination in this material is very low. As the thickness of silicon-based nanostructures reduces below the effective Bohr radius (about 9 nm, in silicon) the quantum efficiency of light emission from this material increases owing to the quantum confinement effect. Relying on this fact, photoemission capability of silicon nanotubes with very thin sidewalls has been demonstrated.

CARBON NANOTUBES

Carbon nanotubes (CNTs) are tubes made of carbon with diameters typically measured in nanometers.

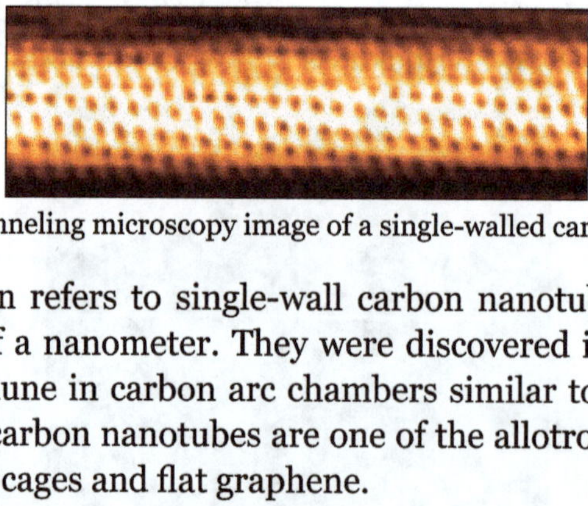

A scanning tunneling microscopy image of a single-walled carbon nanotube.

Carbon nanotubes often refers to single-wall carbon nanotubes (SWCNTs) with diameters in the range of a nanometer. They were discovered independently by Iijima and Ichihashi and Bethune in carbon arc chambers similar to those used to produce fullerenes. Single-wall carbon nanotubes are one of the allotropes of carbon, intermediate between fullerene cages and flat graphene.

Although not made this way, single-wall carbon nanotubes can be thought of as cutouts from a two-dimensional hexagonal lattice of carbon atoms rolled up along one of the Bravais lattice vectors of the hexagonal lattice to form a hollow cylinder. In this construction, periodic boundary conditions are imposed over the length of this roll up vector to yield a lattice with helical symmetry of seamlessly bonded carbon atoms on the cylinder surface.

Carbon nanotubes also often refer to multi-wall carbon nanotubes (MWCNTs) consisting of nested single-wall carbon nanotubes. If not identical, these tubes are very similar to Oberlin, Endo and Koyama's long straight and parallel carbon layers cylindrically rolled around a hollow tube. Multi-wall carbon nanotubes are also sometimes used to refer to double- and triple-wall carbon nanotubes.

Carbon nanotubes can also refer to tubes with an undetermined carbon-wall structure and diameters less than 100 nanometers. Such tubes were discovered by Radushkevich and Lukyanovich. While nanotubes of other compositions exist, most research has been focused on the carbon ones. Therefore, the "carbon" qualifier is often left implicit in the acronyms, and the names are abbreviated NT, SWNT, and MWNT.

Carbon nanotubes can exhibit remarkable electrical conductivity. They also have exceptional tensile strength and thermal conductivity, because of their nanostructure and strength of the bonds between carbon atoms. In addition, they can be chemically modified. These properties are expected to be valuable in many areas of technology, such as electronics, optics, composite materials (replacing or complementing carbon fibers), nanotechnology, and other applications of materials science.

Individual carbon nanotubes naturally align themselves into "ropes" held together by relatively weak van der Waals forces. The length of a carbon nanotube produced by common production methods is often not reported, but is much larger than its diameter. Although rare, nanotubes half a meter long have been created, with a length-to-diameter ratio of more than 100,000,000:1. For many purposes, the length of carbon nanotubes can be assumed to be infinite.

Rolling up a hexagonal lattice along different directions to form different single-wall carbon nanotubes shows that all of these tubes have helical and translational symmetry along the tube axis and many also have nontrivial rotational symmetry about this axis. In addition, most are chiral, meaning the tube and its mirror image cannot be superimposed. This construction also allows single-wall carbon nanotubes to be labeled by a pair of small integers.

A special group of achiral single-wall carbon nanotubes are metallic, but all the rest are either small or moderate band gap semiconductors. These electrical properties, however, do not depend of whether the tube is rolled up above or below the graphene plane and hence are the same for a tube and its mirror image.

Structure of Single-walled Tubes

The structure of an ideal (infinitely long) single-walled carbon nanotube is that of a regular hexagonal lattice drawn on an infinite cylindrical surface, whose vertices are the positions of the carbon atoms. Since the length of the carbon-carbon bonds is fairly fixed, there are constraints on the diameter of the cylinder and the arrangement of the atoms on it.

Zigzag and Armchair Configurations

In the study of nanotubes, one defines a "zigzag" path on a graphene-like lattice as a path that turns 60 degrees, alternating left and right, after stepping through each bond. It is also conventional to define an "armchair" path as one that makes two left turns of 60 degrees followed by two right turns every four steps.

On some carbon nanotubes, there is a closed zigzag path that goes around the tube. One says that the tube is of the zigzag type or configuration, or simply is a zigzag nanotube. If the tube is instead encircled by a closed armchair path, it is said to be of the armchair type, or an armchair nanotube.

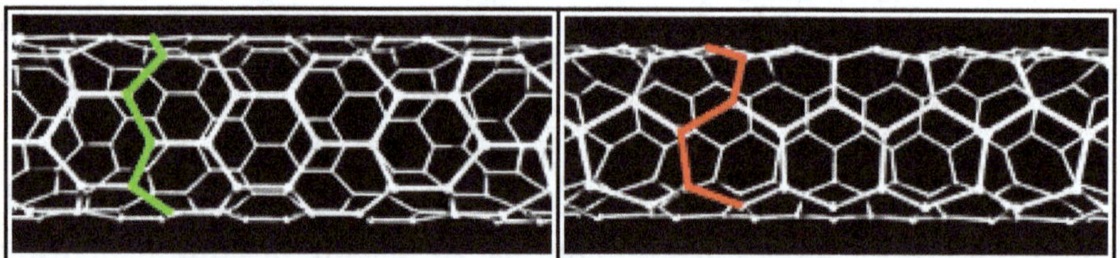

Zigzag nanotube.	Armchair nanotube.

An infinite nanotube that is of the zigzag (or armchair) type consists entirely of closed zigzag (or armchair) paths, connected to each other.

The (*n*,*m*) Notation

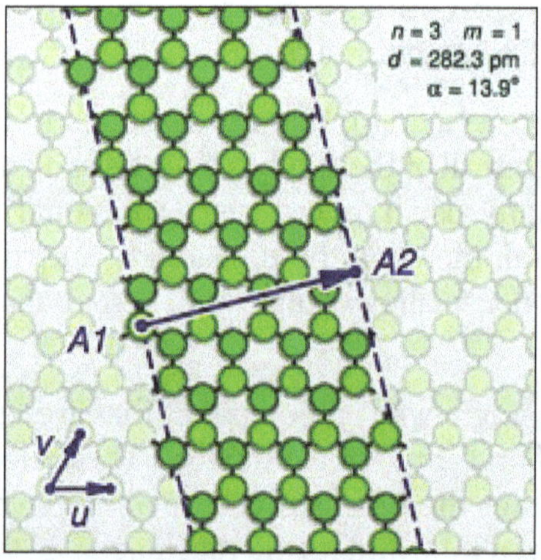

A "sliced and unrolled" representation of a carbon nanotube as a strip of a graphene molecule, overlaid on diagram of the full molecule (faint background). The arrow shows the gap *A2* where the atom *A1* on one edge of the strip would fit in the opposite edge, as the strip is rolled up.

The zigzag and armchair configurations are not the only structures that a single-walled nanotube can have. To describe the structure of a general infinitely long tube, one should imagine it being sliced open by a cut parallel to its axis, that goes through some atom *A*, and then unrolled flat on the plane, so that its atoms and bonds coincide with those of an imaginary graphene sheet—more precisely, with an infinitely long strip of that sheet.

The two halves of the atom *A* will end up on opposite edges of the strip, over two atoms

A1 and *A2* of the graphene. The line from *A1* to *A2* will correspond to the circumference of the cylinder that went through the atom *A*, and will be perpendicular to the edges of the strip.

In the graphene lattice, the atoms can be split into two classes, depending on the directions of their three bonds. Half the atoms have their three bonds directed the same way, and half have their three bonds rotated 180 degrees relative to the first half. The atoms *A1* and *A2*, which correspond to the same atom *A* on the cylinder, must be in the same class.

It follows that the circumference of the tube and the angle of the strip are not arbitrary, because they are constrained to the lengths and directions of the lines that connect pairs of graphene atoms in the same class.

Let *u* and *v* be two linearly independent vectors that connect the graphene atom *A1* to two of its nearest atoms with the same bond directions. That is, if one numbers consecutive carbons around a graphene cell with C1 to C6, then *u* can be the vector from C1 to C3, and *v* be the vector from C1 to C5. Then, for any other atom *A2* with same class as *A1*, the vector from *A1* to *A2* can be written as a linear combination $n\ u + m\ v$, where *n* and *m* are integers. And, conversely, each pair of integers (*n*,*m*) defines a possible position for *A2*.

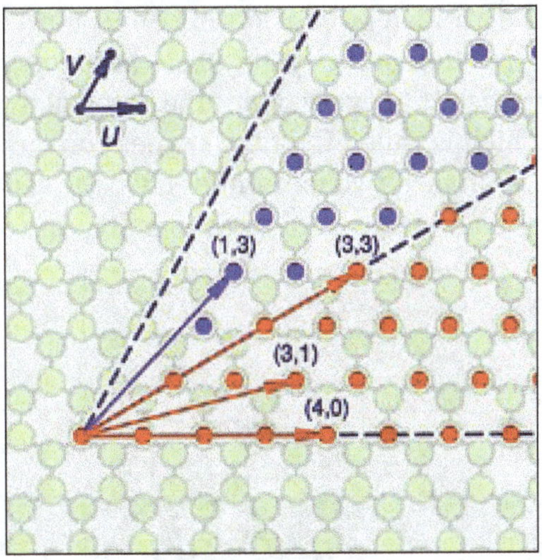

The basis vectors ***u*** and ***v*** of the relevant sub-lattice, the (n,m) pairs that define non-isomorphic carbon nanotube structures (red dots), and the pairs that define the enantiomers of the chiral ones (blue dots).

Given *n* and *m*, one can reverse this theoretical operation by drawing the vector *w* on the graphene lattice, cutting a strip of the latter along lines perpendicular to *w* through its endpoints *A1* and *A2*, and rolling the strip into a cylinder so as to bring those two points together. If this construction is applied to a pair (*k*,0), the result is a zigzag nanotube, with closed zigzag paths of 2*k* atoms. If it is applied to a pair (*k*,*k*), one obtains an armchair tube, with closed armchair paths of 4*k* atoms.

Nanotube Types

Moreover, the structure of the nanotube is not changed if the strip is rotated by 60 degrees clockwise around *A1* before applying the hypothetical reconstruction above. Such a rotation changes the corresponding pair (n,m) to the pair $(-2m,n+m)$.

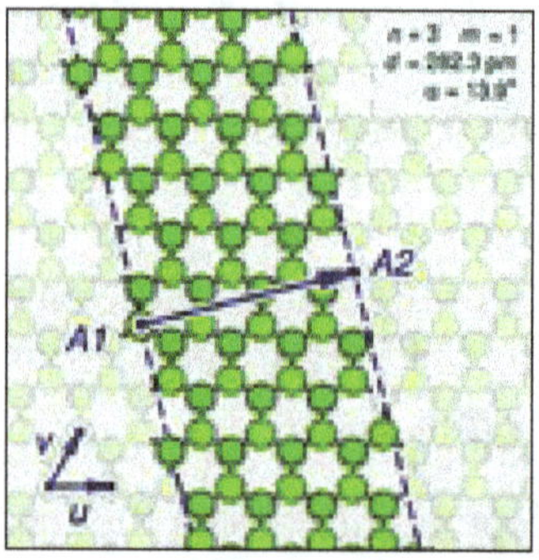

Chiral nanotube of the (3,1) type.

It follows that many possible positions of *A2* relative to *A1* — that is, many pairs (n,m) — correspond to the same arrangement of atoms on the nanotube. That is the case, for example, of the six pairs (1,2), (−2,3), (−3,1), (−1,−2), (2,−3), and (3,−1). In particular, the pairs $(k,0)$ and $(0,k)$ describe the same nanotube geometry.

These redundancies can be avoided by considering only pairs (n,m) such that $n > 0$ and $m \geq 0$; that is, where the direction of the vector w lies between those of u (inclusive) and v (exclusive). It can be verified that every nanotube has exactly one pair (n,m) that satisfies those conditions, which is called the tube's type. Conversely, for every type there is a hypothetical nanotube. In fact, two nanotubes have the same type if and only if one can be conceptually rotated and translated so as to match the other exactly.

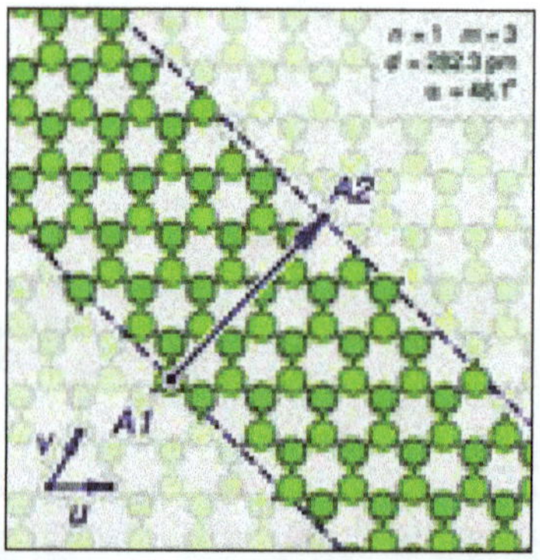

Chiral nanotube of the (1,3) type, mirror image of the (3,1) type.

Instead of the type (n,m), the structure of a carbon nanotube can be specified by giving the length of the vector w (that is, the circumference of the nanotube), and the angle α between the directions of u and w, which may range from 0 (inclusive) to 60 degrees clockwise (exclusive). If the diagram is drawn with u horizontal, the latter is the tilt of the strip away from the vertical.

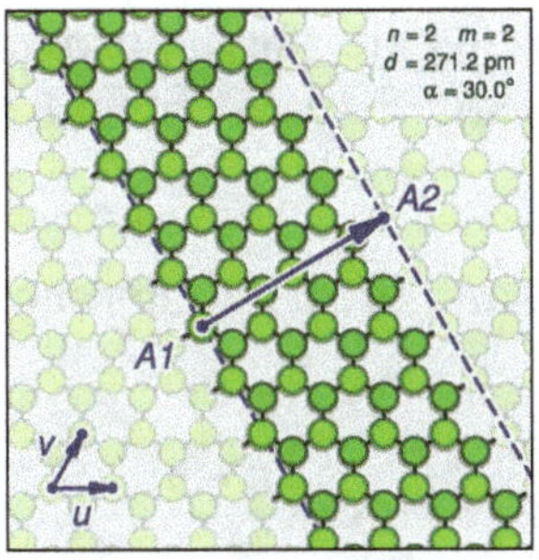

Nanotube of the (2,2) type, the narrowest "armchair" one.

Here are some unrolled nanotube diagrams:

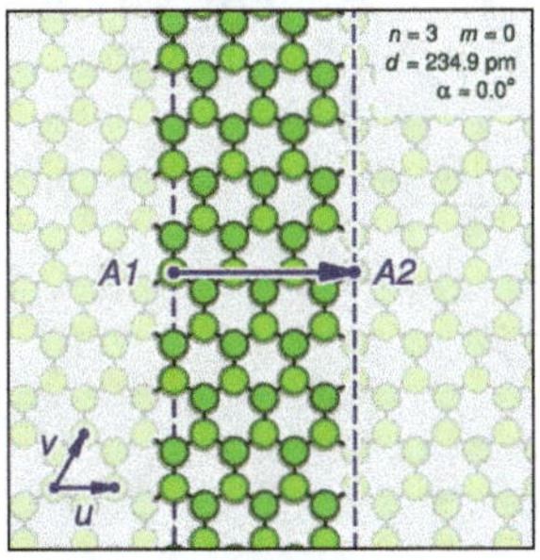

Nanotube of the (3,0) type, the narrowest "zigzag" one.

Chiralty and Mirror Symmetry

A nanotube is chiral if it has type (n,m), with $m > 0$ and $m \neq n$; then its enantiomer (mirror image) has type (m,n), which is different from (n,m). This operation corresponds to mirroring the unrolled strip about the line L through $A1$ that makes an angle of 30 degrees clockwise from the direction of the $\boldsymbol{u}$ vector (that is, with the direction of the vector $\boldsymbol{u}+\boldsymbol{v}$). The only types of nanotubes that are achiral are the $(k,0)$ "zigzag" tubes and the (k,k) "armchair" tubes.

If two enantiomers are to be considered the same structure, then one may consider only types (n,m) with $0 \leq m \leq n$ and $n > 0$. Then the angle α between $\boldsymbol{u}$ and $\boldsymbol{w}$, which may range from 0 to 30 degrees (inclusive both), is called the "chiral angle" of the nanotube.

Circumference and Diameter

From n and m one can also compute the circumference c, which is the length of the vector $\boldsymbol{w}$, which turns out to be:

$$c = |\mathbf{u}|\sqrt{(n^2 + nm + m^2)} \approx 246\sqrt{((n+m)^2 - nm)}$$

in picometres. The diameter d of the tube is then c/π, that is:

$$d \approx 78.3\sqrt{((n+m)^2 - nm)}$$

also in picometres. (These formulas are only approximate, especially for small n and m where the bonds are strained; and they do not take into account the thickness of the wall).

The tilt angle α between $\boldsymbol{u}$ and $\boldsymbol{w}$ and the circumference c are related to the type indices n and m by:

$$\alpha = \arg(n + m/2, m\sqrt{3}/2) = \arc\cos\frac{n + m/2}{c},$$

where arg(x,y) is the clockwise angle between the X-axis and the vector (x,y); a function that is available in many programming languages as atan2(y,x). Conversely, given c and α, one can get the type (n,m) by the formulas:

$$= \frac{2c}{\sqrt{3}}\sin\alpha \qquad n = c\cos\alpha - \frac{m}{2}$$

which must evaluate to integers.

Physical Limits

Narrowest Nanotubes

If n and m are too small, the structure described by the pair (n,m) will describe a molecule that cannot be reasonably called a "tube", and may not even be stable. For example, the structure theoretically described by the pair (1,0) (the limiting "zigzag" type) would be just a chain of carbons. That is a real molecule, the carbyne; which has some characteristics of nanotubes (such as orbital hybridization, high tensile strength, etc.) — but has no hollow space, and may not be obtainable as a condensed phase. The pair (2,0) would theoretically yield a chain of fused 4-cycles; and (1,1), the limiting "armchair" structure, would yield a chain of bi-connected 4-rings. These structures may not be realizable.

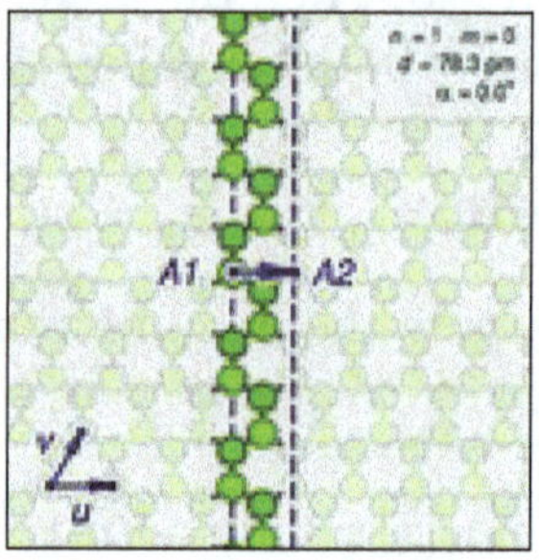

Degenerate "zigzag" tube type (1,0).

The thinnest carbon nanotube proper is the armchair structure with type (2,2), which has a diameter of 0.3 nm. This nanotube was grown inside a multi-walled carbon nanotube. Assigning of the carbon nanotube type was done by a combination of high-resolution transmission electron microscopy (HRTEM), Raman spectroscopy, and density functional theory (DFT) calculations.

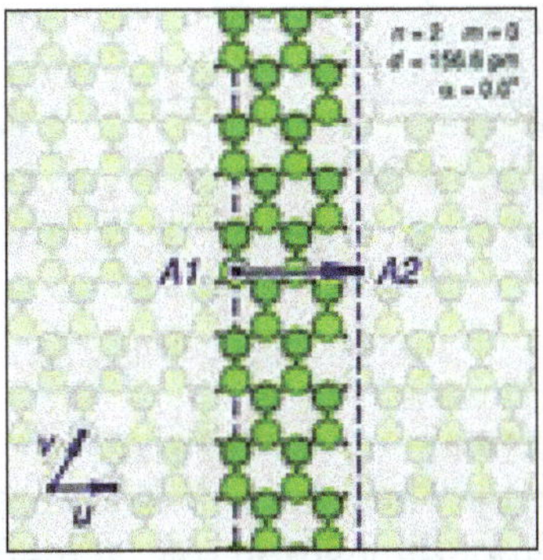

Degenerate "zigzag" tube type (2,0).

The thinnest freestanding single-walled carbon nanotube is about 0.43 nm in diameter. Researchers suggested that it can be either (5,1) or (4,2) SWCNT, but the exact type of the carbon nanotube remains questionable. (3,3), (4,3), and (5,1) carbon nanotubes (all about 0.4 nm in diameter) were unambiguously identified using aberration-corrected high-resolution transmission electron microscopy inside double-walled CNTs.

Here are some tube types that are "degenerate" for being too narrow:

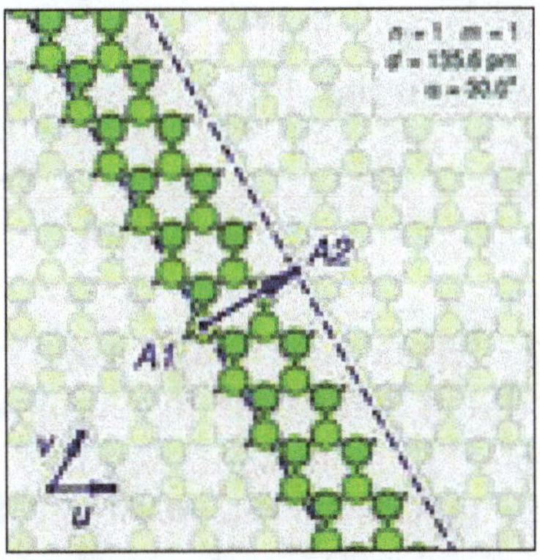

Degenerate "armchair" tube type (1,1).

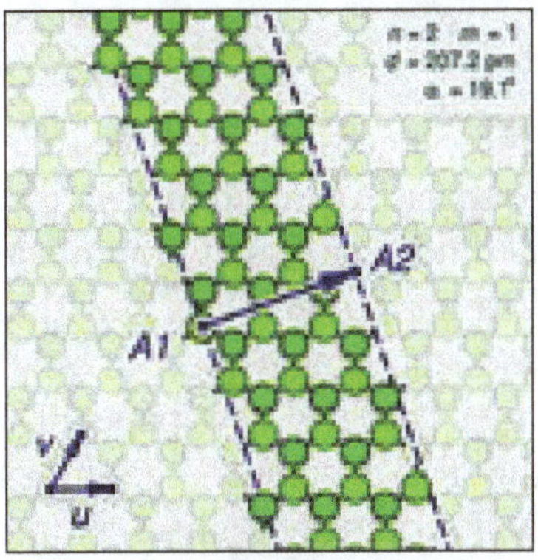

Possibly degenerate chiral tube type (2,1).

Length

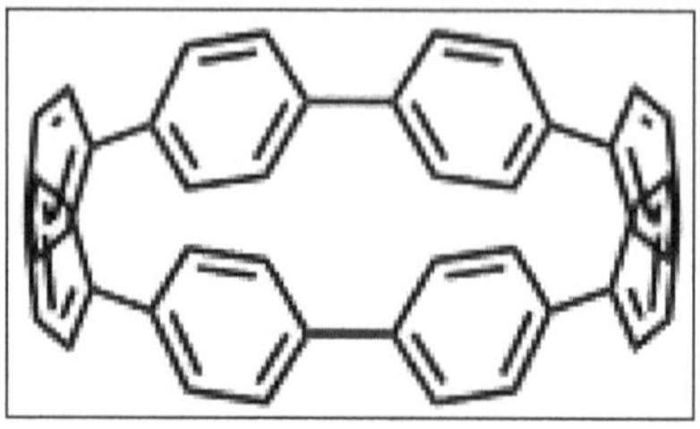

Cycloparaphenylene.

The observation of the *longest* carbon nanotubes grown so far, around 1/2 meter (550 mm long), was reported in 2013. These nanotubes were grown on silicon substrates using an improved chemical vapor deposition (CVD) method and represent electrically uniform arrays of single-walled carbon nanotubes.

The *shortest* carbon nanotube can be considered to be the organic compound cycloparaphenylene, which was synthesized in 2008.

Density

The *highest density* of CNTs was achieved in 2013, grown on a conductive titanium-coated copper surface that was coated with co-catalysts cobalt and molybdenum at lower than typical temperatures of 450 °C. The tubes averaged a height of 380 nm and a mass density of 1.6 g cm^{-3}. The material showed ohmic conductivity (lowest resistance ~22 kΩ).

Variants

There is no consensus on some terms describing carbon nanotubes in scientific literature: both "-wall" and "-walled" are being used in combination with "single", "double", "triple", or "multi", and the letter C is often omitted in the abbreviation, for example, multi-walled carbon nanotube (MWNT). International Standards Organization uses single-wall or multi-wall in its documents.

Multi-walled

Multi-walled nanotubes (MWNTs) consist of multiple rolled layers (concentric tubes) of graphene. There are two models that can be used to describe the structures of multi-walled nanotubes. In the *Russian Doll* model, sheets of graphite are arranged in concentric cylinders, e.g., a (0,8) single-walled nanotube (SWNT) within a larger (0,17) single-walled nanotube. In the *Parchment* model, a single sheet of graphite is rolled in around itself, resembling a scroll of parchment or a rolled newspaper. The interlayer distance in multi-walled nanotubes is close to the distance between graphene layers in graphite, approximately 3.4 Å. The Russian Doll structure is observed more commonly. Its individual shells can be described as SWNTs, which can be metallic or semiconducting. Because of statistical probability and restrictions on the relative diameters of the individual tubes, one of the shells, and thus the whole MWNT, is usually a zero-gap metal.

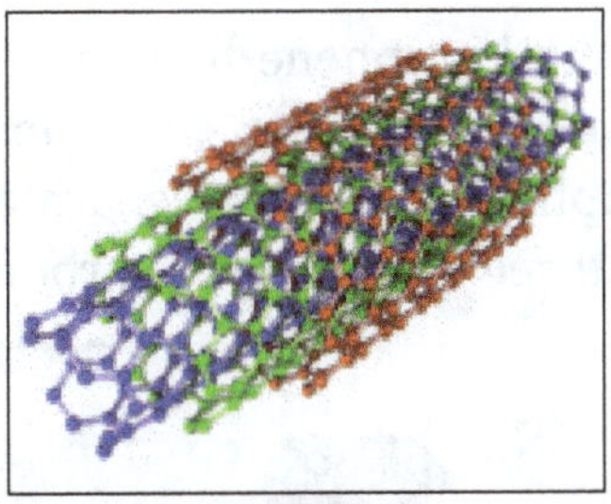

Triple-walled armchair carbon nanotube.

Double-walled carbon nanotubes (DWNTs) form a special class of nanotubes because their morphology and properties are similar to those of SWNTs but they are more resistant to chemicals. This is especially important when it is necessary to graft chemical functions to the surface of the nanotubes (functionalization) to add properties to the CNT. Covalent functionalization of SWNTs will break some C=C double bonds, leaving "holes" in the structure on the nanotube and thus modifying both its mechanical and electrical properties. In the case of DWNTs, only the outer wall is modified. DWNT synthesis on the gram-scale by the CCVD technique was first proposed in 2003 from the selective reduction of oxide solutions in methane and hydrogen.

The telescopic motion ability of inner shells and their unique mechanical properties will permit the use of multi-walled nanotubes as the main movable arms in upcoming nanomechanical devices. The retraction force that occurs to telescopic motion is caused by the Lennard-Jones interaction between shells, and its value is about 1.5 nN.

Junctions and Crosslinking

Transmission electron microscope image of carbon nanotube junction.

Junctions between two or more nanotubes have been widely discussed theoretically. Such junctions are quite frequently observed in samples prepared by arc discharge as well as by chemical vapor deposition. The electronic properties of such junctions were first considered theoretically by Lambin , who pointed out that a connection between a metallic tube and a semiconducting one would represent a nanoscale heterojunction.

Such a junction could therefore form a component of a nanotube-based electronic circuit. The adjacent image shows a junction between two multiwalled nanotubes.

Junctions between nanotubes and graphene have been considered theoretically and studied experimentally. Nanotube-graphene junctions form the basis of pillared graphene, in which parallel graphene sheets are separated by short nanotubes. Pillared graphene represents a class of three-dimensional carbon nanotube architectures.

3D carbon scaffolds.

Recently, several studies have highlighted the prospect of using carbon nanotubes as building blocks to fabricate three-dimensional macroscopic (>100 nm in all three dimensions) all-carbon devices. Lalwani have reported a novel radical-initiated thermal crosslinking method to fabricate macroscopic, free-standing, porous, all-carbon scaffolds using single- and multi-walled carbon nanotubes as building blocks. These scaffolds possess macro-, micro-, and nano-structured pores, and the porosity can be tailored for specific applications. These 3D all-carbon scaffolds/architectures may be used for the fabrication of the next generation of energy storage, supercapacitors, field emission transistors, high-performance catalysis, photovoltaics, and biomedical devices, implants, and sensors.

Other Morphologies

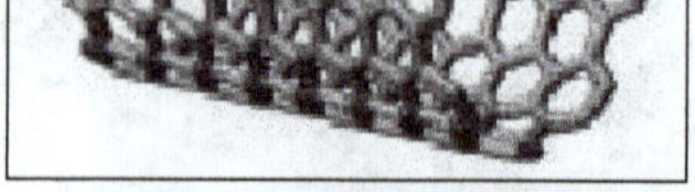

A stable nanobud structure.

Carbon nanobuds are a newly created material combining two previously discovered allotropes of carbon: carbon nanotubes and fullerenes. In this new material, fullerene-like "buds" are covalently bonded to the outer sidewalls of the underlying carbon nanotube. This hybrid material has useful properties of both fullerenes and carbon nanotubes. In particular, they have been found to be exceptionally good field emitters. In composite materials, the attached fullerene molecules may function as molecular anchors preventing slipping of the nanotubes, thus improving the composite's mechanical properties.

A carbon peapod is a novel hybrid carbon material which traps fullerene inside a carbon nanotube. It can possess interesting magnetic properties with heating and irradiation. It can also be applied as an oscillator during theoretical investigations and predictions.

In theory, a nanotorus is a carbon nanotube bent into a torus (doughnut shape). Nanotori are predicted to have many unique properties, such as magnetic moments 1000 times larger than that previously expected for certain specific radii. Properties such as magnetic moment, thermal stability, etc. vary widely depending on the radius of the torus and the radius of the tube.

Graphenated carbon nanotubes are a relatively new hybrid that combines graphitic foliates grown along the sidewalls of multiwalled or bamboo style CNTs. The foliate density can vary as a function of deposition conditions (e.g., temperature and time) with their structure ranging from a few layers of graphene (< 10) to thicker, more graphite-like. The fundamental advantage of an integrated graphene-CNT structure is the high surface area three-dimensional framework of the CNTs coupled with the high edge density of graphene. Depositing a high density of graphene foliates along the length of aligned CNTs can significantly increase the total charge capacity per unit of nominal area as compared to other carbon nanostructures.

Cup-stacked carbon nanotubes (CSCNTs) differ from other quasi-1D carbon structures, which normally behave as quasi-metallic conductors of electrons. CSCNTs exhibit semiconducting behavior because of the stacking microstructure of graphene layers.

Properties

Many properties of single-walled carbon nanotubes depend significantly on the (n,m) type, and this dependence is non-monotonic (see Kataura plot). In particular, the band gap can vary from zero to about 2 eV and the electrical conductivity can show metallic or semiconducting behavior.

Mechanical

Carbon nanotubes are the strongest and stiffest materials yet discovered in terms of tensile strength and elastic modulus. This strength results from the covalent sp^2 bonds formed between the individual carbon atoms. In 2000, a multiwalled carbon nanotube was tested to have a tensile strength of 63 gigapascals (9,100,000 psi). (For illustration, this translates into the ability to endure tension of a weight equivalent to 6,422 kilograms-force (62,980 N; 14,160 lbf) on a cable with cross-section of 1 square millimetre (0.0016 sq in)). Further studies, such as one conducted in 2008, revealed that individual CNT shells have strengths of up to $\approx$100 gigapascals (15,000,000 psi), which is in agreement with quantum/atomistic models. Because carbon nanotubes have a low density for a solid of 1.3 to 1.4 g/cm^3, its specific strength of up to 48,000 kN·m·kg^{-1} is the best of known materials, compared to high-carbon steel's 154 kN·m·kg^{-1}.

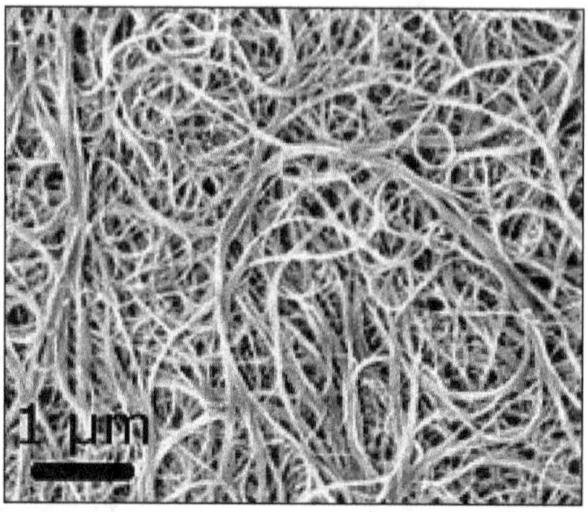

A scanning electron microscopy image of carbon nanotube bundles.

Although the strength of individual CNT shells is extremely high, weak shear interactions between adjacent shells and tubes lead to significant reduction in the effective strength of multiwalled carbon nanotubes and carbon nanotube bundles down to only a few GPa. This limitation has been recently addressed by applying high-energy electron irradiation, which crosslinks inner shells and tubes, and effectively increases the strength of these materials to ≈60 GPa for multiwalled carbon nanotubes and ≈17 GPa for double-walled carbon nanotube bundles. CNTs are not nearly as strong under compression. Because of their hollow structure and high aspect ratio, they tend to undergo buckling when placed under compressive, torsional, or bending stress.

On the other hand, there was evidence that in the radial direction they are rather soft. The first transmission electron microscope observation of radial elasticity suggested that even van der Waals forces can deform two adjacent nanotubes. Later, nanoindentations with an atomic force microscope were performed by several groups to quantitatively measure radial elasticity of multiwalled carbon nanotubes and tapping/contact mode atomic force microscopy was also performed on single-walled carbon nanotubes. Young's modulus of on the order of several GPa showed that CNTs are in fact very soft in the radial direction.

Electrical

Unlike graphene, which is a two-dimensional semimetal, carbon nanotubes are either metallic or semiconducting along the tubular axis. For a given (n,m) nanotube, if $n = m$, the nanotube is metallic; if $n - m$ is a multiple of 3 and $n \neq m$ and $nm \neq 0$, then the nanotube is quasi-metallic with a very small band gap, otherwise the nanotube is a moderate semiconductor. Thus, all armchair ($n = m$) nanotubes are metallic, and nanotubes (6,4), (9,1), etc. are semiconducting. Carbon nanotubes are not semimetallic because the degenerate point (the point where the π [bonding] band meets the π^* [anti-bonding] band, at which the energy goes to zero) is slightly shifted away from the

K point in the Brillouin zone because of the curvature of the tube surface, causing hybridization between the σ^* and π^* anti-bonding bands, modifying the band dispersion.

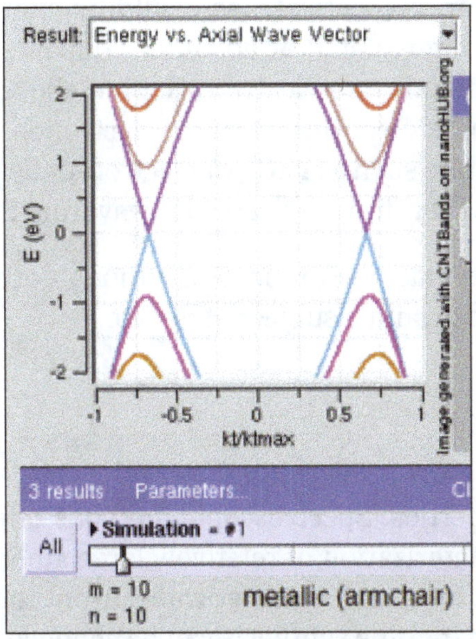

Band structures computed using tight binding approximation for (6,0) CNT (zigzag, metallic), (10,2) CNT (semiconducting) and (10,10) CNT (armchair, metallic).

The rule regarding metallic versus semiconductor behavior has exceptions because curvature effects in small-diameter tubes can strongly influence electrical properties. Thus, a (5,0) SWCNT that should be semiconducting in fact is metallic according to the calculations. Likewise, zigzag and chiral SWCNTs with small diameters that should be metallic have a finite gap (armchair nanotubes remain metallic). In theory, metallic nanotubes can carry an electric current density of 4×10^9 A/cm^2, which is more than 1,000 times greater than those of metals such as copper, where for copper interconnects, current densities are limited by electromigration. Carbon nanotubes are thus being explored as interconnects and conductivity-enhancing components in composite materials, and many groups are attempting to commercialize highly conducting electrical wire assembled from individual carbon nanotubes. There are significant challenges to be overcome however, such as undesired current saturation under voltage, and the much more resistive nanotube-to-nanotube junctions and impurities, all of which lower the electrical conductivity of the macroscopic nanotube wires by orders of magnitude, as compared to the conductivity of the individual nanotubes.

Because of its nanoscale cross-section, electrons propagate only along the tube's axis. As a result, carbon nanotubes are frequently referred to as one-dimensional conductors. The maximum electrical conductance of a single-walled carbon nanotube is $2G_0$, where $G_0 = 2e^2/h$ is the conductance of a single ballistic quantum channel.

Because of the role of the π-electron system in determining the electronic properties of graphene, doping in carbon nanotubes differs from that of bulk crystalline semiconductors from the same group of the periodic table (e.g., silicon). Graphitic substitution of carbon atoms in the nanotube wall by boron or nitrogen dopants leads to p-type and n-type behavior, respectively, as would be expected in silicon. However, some non-substitutional (intercalated or adsorbed) dopants introduced into a carbon nanotube, such as alkali metals and electron-rich metallocenes, result in n-type conduction because they donate electrons to the π-electron system of the nanotube. By contrast, π-electron acceptors such as $FeCl_3$ or electron-deficient metallocenes function as p-type dopants because they draw π-electrons away from the top of the valence band.

Intrinsic superconductivity has been reported, although other experiments found no evidence of this, leaving the claim a subject of debate.

Optical

Carbon nanotubes have useful absorption, photoluminescence (fluorescence), and Raman spectroscopy properties. Spectroscopic methods offer the possibility of quick and non-destructive characterization of relatively large amounts of carbon nanotubes. There is a strong demand for such characterization from the industrial point of view: numerous parameters of nanotube synthesis can be changed, intentionally or unintentionally, to alter the nanotube quality. As shown below, optical absorption, photoluminescence, and Raman spectroscopies allow quick and reliable characterization of this "nanotube quality" in terms of non-tubular carbon content, structure (chirality) of the produced nanotubes, and structural defects. These features determine nearly any other properties such as optical, mechanical, and electrical properties.

Carbon nanotubes are unique "one-dimensional systems" which can be envisioned as rolled single sheets of graphite (or more precisely graphene). This rolling can be done at different angles and curvatures resulting in different nanotube properties. The diameter typically varies in the range 0.4–40 nm (i.e., "only" ~100 times), but the length can vary ~100,000,000,000 times, from 0.14 nm to 55.5 cm. The nanotube aspect ratio, or the length-to-diameter ratio, can be as high as 132,000,000:1, which is unequalled by any other material. Consequently, all the properties of the carbon nanotubes relative to those of typical semiconductors are extremely anisotropic (directionally dependent) and tunable.

Whereas mechanical, electrical, and electrochemical (supercapacitor) properties of the carbon nanotubes are well established and have immediate applications, the practical use of optical properties is yet unclear. The aforementioned tunability of properties is potentially useful in optics and photonics. In particular, light-emitting diodes (LEDs) and photo-detectors based on a single nanotube have been produced in the lab. Their unique feature is not the efficiency, which is yet relatively low, but the narrow selectivity in the wavelength of emission and detection of light and the possibility of its fine tuning through the nanotube structure. In addition, bolometer and optoelectronic memory devices have been realised on ensembles of single-walled carbon nanotubes.

Crystallographic defects also affect the tube's electrical properties. A common result is lowered conductivity through the defective region of the tube. A defect in armchair-type tubes (which can conduct electricity) can cause the surrounding region to become semi-conducting, and single monatomic vacancies induce magnetic properties.

Thermal

All nanotubes are expected to be very good thermal conductors along the tube, exhibiting a property known as "ballistic conduction", but good insulators lateral to the tube axis. Measurements show that an individual SWNT has a room-temperature thermal conductivity along its axis of about 3500 $W \cdot m^{-1} \cdot K^{-1}$; compare this to copper, a metal well known for its good thermal conductivity, which transmits 385 $W \cdot m^{-1} \cdot K^{-1}$. An individual SWNT has a room-temperature thermal conductivity across its axis (in the radial direction) of about 1.52 $W \cdot m^{-1} \cdot K^{-1}$, which is about as thermally conductive as soil. Macroscopic assemblies of nanotubes such as films or fibres have reached up to 1500 $W \cdot m^{-1} \cdot K^{-1}$ so far. Networks composed of nanotubes demonstrate different values of thermal conductivity, from the level of thermal insulation with the thermal conductivity of 0.1 $W \cdot m^{-1} \cdot K^{-1}$ to such high values. That is dependent on the amount of contribution to the thermal resistance of the system caused by the presence of impurities, misalignments and other factors. The temperature stability of carbon nanotubes is estimated to be up to 2800 °C in vacuum and about 750 °C in air.

Crystallographic defects strongly affect the tube's thermal properties. Such defects lead to phonon scattering, which in turn increases the relaxation rate of the phonons. This reduces the mean free path and reduces the thermal conductivity of nanotube structures. Phonon transport simulations indicate that substitutional defects such as nitrogen or boron will primarily lead to scattering of high-frequency optical phonons. However, larger-scale defects such as Stone Wales defects cause phonon scattering over a wide range of frequencies, leading to a greater reduction in thermal conductivity.

Modeling

Carbon nanotubes are modelled in a similar manner as traditional composites in which a reinforcement phase is surrounded by a matrix phase. Ideal models such as cylindrical, hexagonal and square models are common. The size of the micromechanics model is highly function of the studied mechanical properties. The concept of representative volume element (RVE) is used to determine the appropriate size and configuration of computer model to replicate the actual behavior of CNT reinforced nanocomposite. Depending on the material property of interest (thermal, electrical, modulus, creep), one RVE might predict the property better than the alternatives. While the implementation of ideal model is comutationally efficient, they do not represent microstructural features observed in scanning electron microscopy of actual nanocomposites. To incorporate realistic modeling, computer models are also generated to incorporate variability such as waviness, orientation and agglomeration of multiwall or single wall carbon nanotubes.

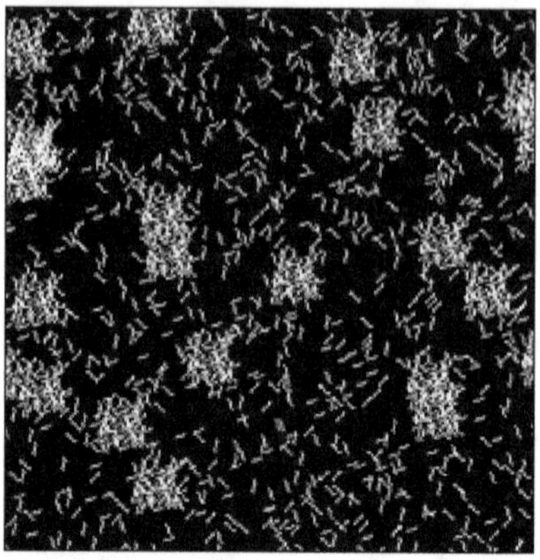

Computer simulated microstructures with agglomeration regions.

Metrology

There are many metrology standards and reference materials available for carbon nanotubes.

For single-wall carbon nanotubes, ISO/TS 10868 describes a measurement method for the diameter, purity, and fraction of metallic nanotubes through optical absorption spectroscopy, while ISO/TS 10797 and ISO/TS 10798 establish methods to characterize the morphology and elemental composition of single-wall carbon nanotubes, using transmission electron microscopy and scanning electron microscopy respectively, coupled with energy dispersive X-ray spectrometry analysis.

NIST SRM 2483 is a soot of single-wall carbon nanotubes used as a reference material for elemental analysis, and was characterized using thermogravimetric analysis, prompt gamma activation analysis, induced neutron activation analysis, inductively coupled plasma mass spectroscopy, resonant Raman scattering, UV-visible-near infrared fluorescence spectroscopy and absorption spectroscopy, scanning electron microscopy, and transmission electron microscopy. The Canadian National Research Council also offers a certified reference material SWCNT-1 for elemental analysis using neutron activation analysis and inductively coupled plasma mass spectroscopy. NIST RM 8281 is a mixture of three lengths of single-wall carbon nanotube.

For multiwall carbon nanotubes, ISO/TR 10929 identifies the basic properties and the content of impurities, while ISO/TS 11888 describes morphology using scanning electron microscopy, transmission electron microscopy, viscometry, and light scattering analysis. ISO/TS 10798 is also valid for multiwall carbon nanotubes.

Chemical Modification

Carbon nanotubes can be functionalized to attain desired properties that can be used in a wide variety of applications. The two main methods of carbon nanotube functionalization are covalent and non-covalent modifications. Because of their apparent hydrophobic nature, carbon nanotubes tend to agglomerate hindering their dispersion in solvents or viscous polymer melts. The resulting nanotube bundles or aggregates reduce the mechanical performance of the final composite. The surface of the carbon nanotubes can be modified to reduce the hydrophobicity and improve interfacial adhesion to a bulk polymer through chemical attachment.

Also surface of carbon nanotubes can be fluorinated or halofluorinated by CVD-method with fluorocarbons, hydro- or halofluorocarbons by heating while in contact of such carbon material with fluoroorganic substance to form partially fluorinated carbons (so called Fluocar materials) with grafted (halo)fluoroalkyl functionality.

Applications

A primary obstacle for applications of carbon nanotubes has been their cost. Prices for single-walled nanotubes declined from around $1500 per gram as of 2000 to retail prices of around $50 per gram of as-produced 40–60% by weight SWNTs as of March 2010. As of 2016, the retail price of as-produced 75% by weight SWNTs was $2 per gram. SWNTs are forecast to make a large impact in electronics applications by 2020 according to *The Global Market for Carbon Nanotubes* report.

Current

Current use and application of nanotubes has mostly been limited to the use of bulk nanotubes, which is a mass of rather unorganized fragments of nanotubes. Bulk nanotube materials may never achieve a tensile strength similar to that of individual tubes, but such composites may, nevertheless, yield strengths sufficient for many applications. Bulk carbon nanotubes have already been used as composite fibers in polymers to improve the mechanical, thermal and electrical properties of the bulk product.

- Easton-Bell Sports, Inc. have been in partnership with Zyvex Performance Materials, using CNT technology in a number of their bicycle components – including flat and riser handlebars, cranks, forks, seatposts, stems and aero bars.
- Amroy Europe Oy manufactures Hybtonite carbon nanoepoxy resins where carbon nanotubes have been chemically activated to bond to epoxy, resulting in a composite material that is 20% to 30% stronger than other composite materials. It has been used for wind turbines, marine paints and a variety of sports gear such as skis, ice hockey sticks, baseball bats, hunting arrows, and surfboards.
- Surrey NanoSystems synthesises carbon nanotubes to create vantablack.

Other current applications include:

- Tips for atomic force microscope probes.
- In tissue engineering, carbon nanotubes can act as scaffolding for bone growth.

Under Development

Current research for modern applications include:

- Using carbon nanotubes as a scaffold for diverse microfabrication techniques.
- Energy dissipation in self-organized nanostructures under influence of an electric field.
- Using carbon nanotubes for environmental monitoring due to their active surface area and their ability to absorb gases.
- Jack Andraka used carbon nanotubes in his pancreatic cancer test. His method of testing won the Intel International Science and Engineering Fair Gordon E. Moore Award in the spring of 2012.Jack Andraka, the Teen Prodigy of Pancreatic Cancer
- The Boeing Company has patented the use of carbon nanotubes for structural health monitoring of composites used in aircraft structures. This technology will greatly reduce the risk of an in-flight failure caused by structural degradation of aircraft.
- Zyvex Technologies has also built a 54' maritime vessel, the Piranha Unmanned Surface Vessel, as a technology demonstrator for what is possible using CNT technology. CNTs help improve the structural performance of the vessel, resulting in a lightweight 8,000 lb boat that can carry a payload of 15,000 lb over a range of 2,500 miles.

Carbon nanotubes can serve as additives to various structural materials. For instance, nanotubes form a tiny portion of the material(s) in some (primarily carbon fiber) baseball bats, golf clubs, car parts, or damascus steel.

IBM expected carbon nanotube transistors to be used on Integrated Circuits by 2020.

Potential

The strength and flexibility of carbon nanotubes makes them of potential use in controlling other nanoscale structures, which suggests they will have an important role in nanotechnology engineering. The highest tensile strength of an individual multi-walled carbon nanotube has been tested to be 63 GPa. Carbon nanotubes were found in Damascus steel from the 17th century, possibly helping to account for the legendary strength of the swords made of it. Recently, several studies have highlighted the prospect of using carbon nanotubes as building blocks to fabricate three-dimensional macroscopic (>1mm in all three dimensions) all-carbon devices. Lalwani have reported a novel radical initiated thermal crosslinking method to fabricated macroscopic, free-standing, porous, all-carbon scaffolds using single- and multi-walled carbon nanotubes as building blocks. These scaffolds possess macro-, micro-, and nano- structured pores and the porosity can be tailored for specific applications. These 3D all-carbon scaffolds/architectures may be used for the fabrication of the next generation of energy storage, supercapacitors, field emission transistors, high-performance catalysis, photovoltaics, and biomedical devices and implants.

CNTs are potential candidates for future via and wire material in nano-scale VLSI circuits. Eliminating electromigration reliability concerns that plague today's Cu interconnects, isolated (single and multi-wall) CNTs can carry current densities in excess of 1000 MA/cm^2 without electromigration damage.

Single-walled nanotubes are likely candidates for miniaturizing electronics. The most basic building block of these systems is an electric wire, and SWNTs with diameters of an order of a nanometer can be excellent conductors. One useful application of SWNTs is in the development of the first intermolecular field-effect transistors (FET). The first intermolecular logic gate using SWCNT FETs was made in 2001. A logic gate requires both a p-FET and an n-FET. Because SWNTs are p-FETs when exposed to oxygen and n-FETs otherwise, it is possible to expose half of an SWNT to oxygen and protect the other half from it. The resulting SWNT acts as a *not* logic gate with both p- and n-type FETs in the same molecule. Large quantities of pure CNTs can be made into a free-standing sheet or film by surface-engineered tape-casting (SETC) fabrication technique which is a scalable method to fabricate flexible and foldable sheets with superior properties. Another reported form factor is CNT fiber (a.k.a. filament) by wet spinning. The fiber is either directly spun from the synthesis pot or spun from pre-made dissolved CNTs. Individual fibers can be turned into a yarn. Apart from its strength and flexibility, the main advantage is making an electrically conducting yarn. The electronic properties of individual CNT fibers (i.e. bundle of individual CNT) are governed by the two-dimensional structure of CNTs. The fibers were measured to have a resistivity only one order of magnitude higher than metallic conductors at 300K. By further optimizing the CNTs and CNT fibers, CNT fibers with improved electrical properties could be developed.

CNT-based yarns are suitable for applications in energy and electrochemical water treatment when coated with an ion-exchange membrane. Also, CNT-based yarns could replace copper as a winding material.

Applications of Carbon Nanotubes

The unique nature of carbon combines with the molecular perfection of single-wall CNTs to endow them with extraordinary material properties, such as very high thermal and electrical conductivity, stiffness, strength, and toughness. It is the only element in the periodic table which bonds to itself in an extended network with the strength of the carbon-carbon bond. The delocalized pi-electron donated by each atom is free to move about the whole structure, instead of remaining with its donor atom, resulting in the first known molecule with metallic-type electrical conductivity. Moreover, an intrinsic thermal conductivity higher than even diamond is offered by the high-frequency carbon-carbon bond vibrations.

In most materials, however, due to the occurrence of defects in their structure, the actual observed material properties such as strength, electrical conductivity, and so on are degraded very significantly. For example, high-strength steel typically fails at only around 1% of its theoretical breaking strength. However, CNTs achieve values very near to their theoretical limits owing to their molecular perfection of structure. This aspect is part of the unique story of CNTs. CNTs are examples of true nanotechnology: they are only about a nanometer in diameter, but are molecules that can be manipulated physically and chemically in very useful ways. They find an incredible range of applications in electronics, materials science, energy management, chemical processing, and many other fields.

CNTs Thermal Conductivity

CNTs have outstanding heat conductivity, electrical conductivity, and mechanical properties. They are probably the best electron field-emitter possible. They are polymers of pure carbon and can be made to and manipulated using the recognized and extremely rich chemistry of carbon. This offers the opportunity to alter their structure and to optimize their dispersion and solubility. Most notably, CNTs are molecularly perfect, in the sense that they are generally free of property-degrading flaws in the nanotube structure. Their material properties can thus reach close to the very high levels intrinsic to them. Due to these extraordinary characteristics, CNTs can be prospectively used in a number of applications.

CNTs Field Emission Applications

CNTs are the best known field emitters of any material. This is understandable, with regard to their high electrical conductivity, and the unbelievable sharpness of their tip (as the tip's radius of curvature becomes smaller, the electric field will be more concentrated, resulting in increased field emission; this is the same reason lightning rods are sharp). In addition, the sharpness of the tip also indicates that they emit at specifically low voltage, a key fact for building low-power electrical devices that employ this feature. CNTs can carry an amazingly high current density, probably as high as 10^{13} A/cm^2. Additionally, the current is extremely stable. Field-emission flat-panel displays are an immediate application of this behavior, receiving considerable interest. Unlike conventional cathode ray tube display where a single electron gun is used, CNT-based displays use a separate electron gun (or even many of them) for each individual pixel in the display. Their low turn-on and operating voltages, high current density, and steady, long-lived behavior make CNTs very attractive field emitters in this application. General types of low-voltage cold-cathode lighting sources, electron microscope sources, and lightning arrestors are other applications utilizing the field-emission characteristics of CNTs.

CNTs Conductive Plastics

Over the past five decades, much of the history of plastics has involved their use as a substitute for metals. For structural applications, plastics have progressed tremendously, but not where electrical conductivity is needed, since plastics are very good electrical insulators. This deficiency can be ruled out by loading plastics up with conductive fillers, such as carbon black and larger graphite fibers (the ones used to make golf clubs and tennis rackets). In order to offer the necessary conductivity using conventional fillers, the loading required is typically high, however, leading to heavy parts and, more prominently, plastic parts whose structural properties are highly degraded. It is well known that as the aspect ratio of filler particles becomes high, the loading required to achieve a given level of conductivity becomes low. For this reason, CNTs are perfect because they have the highest aspect ratio of any carbon fiber. Furthermore, their natural tendency to form ropes offers inherently very long conductive pathways even at ultra-low loadings.

This behavior of CNTs is utilized in applications such as electrostatic dissipation (ESD); EMI/RFI shielding composites; coatings for gaskets, enclosures, and other uses; radar-absorbing materials for low-observable ("stealth") applications; and antistatic materials and (even transparent!) conductive coatings.

CNTs Energy Storage

The intrinsic properties of CNTs make them the preferred material for use as electrodes in capacitors and batteries — two technologies of fast-growing significance. CNTs possess good electrical conductivity, an extremely high surface area (~1000 m^2/g), and most importantly, their linear geometry makes their surface very accessible to the electrolyte.

Research has demonstrated that CNTs have the highest reversible capacity of any carbon material for use in lithium-ion batteries. Moreover, CNTs are excellent materials for supercapacitor electrodes and are currently being marketed for this application.

In addition, CNTs hold applications in various fuel cell components. They have several properties, such as high thermal conductivity and surface area, making them valuable as electrode catalyst supports in PEM fuel cells. Owing to their high electrical conductivity, they may also be used in gas diffusion layers, besides current collectors. The high strength and toughness-to-weight characteristics of CNTs may also prove useful as part of composite components in fuel cells that are used in transport applications, where durability is paramount.

CNTs Conductive Adhesives and Connectors

The exact properties that make CNTs desirable as conductive fillers for use in ESD materials, electromagnetic shielding, and so on make them suitable for interconnection applications and electronics packaging, including coaxial cables, potting compounds, and adhesives and other types of connectors.

CNTs Molecular Electronics

The idea of building electronic circuits out of the critical building blocks of materials – molecules – has seen growth in the past five years, and is a vital part of nanotechnology. In any electronic circuit, but specifically when dimensions reduce in size to the nanoscale, the interconnections between switches and other active devices become more and more essential. Their ability to be precisely derived, electrical conductivity, and geometry make CNTs the most suitable candidates for the connections in molecular electronics. Furthermore, they have been shown as switches themselves.

CNTs Thermal Materials

The record-setting anisotropic thermal conductivity of CNTs is opening doors to several applications that involve heat transfer. Such an application is found in electronics, specifically advanced computing, where uncooled chips currently regularly exceed 100 °C.

The technology for creating aligned structures and ribbons of CNTs is a step toward achieving extremely efficient heat conduits. Furthermore, composites with CNTs have been demonstrated to significantly increase their bulk thermal conductivity, even at incredibly small loadings.

CNTs Structural Composites

The superior properties of CNTs are not just restricted to thermal and electrical conductivities but also include mechanical properties, such as strength, toughness, and stiffness. These properties pave the way for use in a range of applications exploiting them, including advanced composites that need high values of one or more of these properties.

CNTs Fibers and Fabrics

Recently, fibers spun from pure CNTs have been demonstrated and are experiencing rapid development, together with CNT composite fibers. Such super strong fibers will have several applications such as woven fabrics and textiles, transmission line cables, and body and vehicle armor. CNTs are also being employed in order to make textiles stain resistant.

CNT Catalyst Supports

CNTs intrinsically possess an enormously high surface area; actually, for SWNTs, every atom is not just on one surface – but two surfaces, the interior and exterior of the nanotube. Along with the ability to attach basically any chemical species to their sidewalls (functionalization) offers a prospect for unique catalyst supports. Their electrical conductivity may also be used propitiously in the quest for new catalysts and catalytic behavior.

CNTs Biomedical Applications

Although the exploration of CNTs in biomedical applications is just in progress, it has great potential. Since a great part of the human body is made up of carbon, it is usually considered a very biocompatible material. The growth of cells on CNTs has been demonstrated; therefore, they apparently have no toxic effect. The cells also do not adhere to the CNTs, opening doors for applications such as anti-fouling coatings for ships and coatings for prosthetics.

The ability to functionalize (chemically modify) the sidewalls of CNTs also gives rise to biomedical applications including neuron growth and regeneration, and vascular stents. It has also been demonstrated that a single strand of DNA can be bonded to a nanotube, which can subsequently be effectively inserted into a cell.

CNTs Air and Water Filtration

Several corporations and researchers have already developed CNT-based water and air filtration devices. It has been reported that these filters, apart from blocking the tiniest particles, can also destroy most bacteria. This is one more area where CNTs have already been commercialized and products are available now.

CNTs Ceramic Applications

Materials scientists at UC Davis have produced a ceramic material reinforced with carbon nanotubes. The new material is significantly tougher than traditional ceramics, conducts electricity, and can both conduct heat and function as a thermal barrier, with respect to the nanotube orientation.

Since ceramic materials are very hard and resistant to heat and chemical attack, they are valuable for applications such as coating turbine blades; however, they are also very brittle. The researchers mixed powdered alumina (aluminum oxide) with 5%–10% carbon nanotubes, in addition to 5% finely milled niobium. The mixture was treated with an electrical pulse in a process called spark-plasma sintering by the researchers. This process collates ceramic powders more rapidly and at lower temperatures than traditional processes.

The fracture toughness (resistance to cracking under stress) of the new material is up to five times of that of traditional alumina. The material exhibits electrical conductivity seven times of that of earlier ceramics made with nanotubes. It also has fascinating thermal properties, conducting heat in one direction, along the alignment of the nanotubes and, on the other hand, reflecting heat at right angles to the nanotubes, making it a preferred material for thermal barrier coatings.

Other Carbon Nanotubes Applications

There are several other potential applications for CNTs, including solar collection, nanoporous filters, catalyst supports, and all kinds of coatings. There are almost certainly several surprising applications for this excellent material that will be revealed in the future, and which may prove to be the most significant and valuable ones of all. A number of researchers have been studying the conductive and/or waterproof paper produced using CNTs. CNTs have also been demonstrated to absorb infrared light and may hold applications in the I/R optics industry.

References

- P.M. Ajayan; L.S. Schadler; P.V. Braun (2003). Nanocomposite science and technology. Wiley. ISBN 978-3-527-30359-5
- Batista, Carlos A. Silvera; Larson, Ronald G.; Kotov, Nicholas A. (9 October 2015). "Nonadditivity of nanoparticle interactions". Science. 350 (6257): 1242477. doi:10.1126/science.1242477. ISSN 0036-8075. PMID 26450215
- Kotov, Nicholas A. (14 December 2017). "Self-assembly of inorganic nanoparticles: Ab ovo". Europhysics Letters. 119 (6): 66008. Bibcode:2017EL....11966008K. doi:10.1209/0295-5075/119/66008
- Hotze, Ernest M.; Phenrat, Tanapon; Lowry, Gregory V. (2010-11-01). "Nanoparticle Aggregation: Challenges to Understanding Transport and Reactivity in the Environment". Journal of Environment Quality. 39 (6): 1909. doi:10.2134/jeq2009.0462. ISSN 1537-2537
- Nanoparticles as drug delivery systems,Pharmacological Reports, Volume 64, Issue 5,2012,Pages 1020-1037,ISSN 1734-1140, https://doi.org/10.1016/S1734-1140(12)70901-5
- Huang, Xuezhen; Gonzalez-Rodriguez, Roberto; Rich, Ryan; Gryczynski, Zygmunt; Coffer, Jeffery L. (2013). "Fabrication and size dependent properties of porous silicon nanotube arrays". Chemical Communications. 49 (51): 5760. doi:10.1039/C3CC41913D. PMID 23695426

4

Nanoelectronics

The utilization of nanotechnology in electronic components is referred to as nanoelectronics. Various components of nanoelectronics are nanoradios, nanomotors, nanoionic devices, nanobatteries, nanogenerators, nanosensors and nanowires. All these components associated with nanoelectronics have been carefully analyzed in this chapter.

Nanoelectronics is based on the application of nanotechnology in the field of electronics and electronic components. Although the term Nanoelectronics may generally mean all the electronic components, special attention is given in the case of transistors. These transistors have a size lesser than 100 nanometres. Visibly, they are very small that separate studies have to be made for knowing the quantum mechanical properties and inter-atomic design. As a result, though the transistors appear in the nanometre range, they are designed through nanotechnology. Their design is also very much different from the traditional transistors and usually falls in the category of one-dimensional nanotubes/nanowires, hybrid molecular electronics, or advanced molecular electronics.

This technology is said to be the next future, but its practicality is near to impossible even now that they may be difficult to emerge soon.

Although a nanoelectronic device can be made fully functional, the work load it can do is restricted to its size. The basic principle is that the power of a machine will increase according to the increase in volume, but the amount of friction that the machine's bearings hold will depend on the surface area of the machine.

For the small size of the nanoelectronic device cannot be used for the moving of heavy load like a mechanical device. If such a task is tried, it will fail as the available power will be easily overcome by the frictional forces. So, it is sure that these devices have limitations in real world applications.

Different Approaches to Nanoelectronics

- Nanofabrication: This method is used to design arrays or layers of nanoelectronic device to work for a single operation. Nanoelectromechanical systems are also a part of nanofabrication.
- Nanomaterials electronics: In Nanoelectronics, the transistors are packed as arrays on to a single chip. Thus they remain in a uniform manner and symmetrical

in nature. Thus they are known to have a more speedy movement of electrons in the material. The dielectric constant of the device also increases and the electron or hole characteristics also become symmetrical in nature.

Applications of Nanoelectronics

Some of the devices that have been developed with the help of Nanoelectronics and its future applications are listed below:

- Nanoradio.
- Nanocomputers.

The conventional computers with a big processor will be replaced with Nanocomputers with nanoprocessors that will have higher performance and speed than the conventional computers. Researchers are performing various experiments on by using nanolithographic methods to design better nanoprocessors. Experiments are also taking place by replacing the CMOS components in conventional processors with nanowires. The FET's in the computers are replaced by carbon nanotubes.

- Energy production.

The devices using Nanoelectronics technology also includes solar cells that are highly efficient and cheaper than the conventional ones. If such efficient solar energy can be created it would be a revolution to the global energy needs.

Using the technology, researchers are developing a generator for energy production in vivo called bio-nano generators. Basically, the generator is an electrochemical device which is designed in nanoscale size. It works like a fuel cell which generates the power by absorbing the blood glucose in a living body. The glucose will be separated from the body with the help of an enzyme. This enzyme separates the glucose from the electrons and makes them useful for generating power.

The power generated through such a device will be only a few watts as the body itself needs some glucose for its normal functioning. This small power can be used to power up devices placed inside the body like pacemakers or sugar-fed nanorobots.

MOLECULAR WIRE

Molecular wires (or sometimes called molecular nanowires) are molecular chains that conduct electric current. They are the proposed building blocks for molecular electronic devices. Their typical diameters are less than three nanometers, while their lengths may be macroscopic, extending to centimeters or more.

Most types of molecular wires are derived from organic molecules. One naturally occurring molecular wire is DNA. Prominent inorganic examples include polymeric materials such as $Li_2Mo_6Se_6$ and $Mo_6S_{9-x}I_x$, $[Pd_4(CO)_4(OAc)_4Pd(acac)_2]$, and single-molecule extended metal atom chains (EMACs) which comprise strings of transition metal atoms directly bonded to each other. Molecular wires containing paramagnetic inorganic moieties can exhibit Kondo peaks.

Structure

The structure of a $Mo_6S_{9-x}I_x$ molecular wire. Mo atoms are blue, iodine atoms are red and sulphur atoms are yellow.

Unlike the more usual nanowires (which are very thin crystals), molecular nanowires are composed of repeating monomers. In the case of DNA, the repeat units are the nucleotides with a backbone made of sugars and phosphate groups joined by ester bonds. Attached to each sugar is one of four types of bases. In case of $Mo_6S_{9-x}I_x$, the repeat units are $Mo_6S_{9-x}I_x$ clusters, which are joined together by flexible sulfur or iodine bridges. Molecular nanowires can be manipulated and investigated as single molecules, but they often aggregate in solution into swatches or bundles. In the case of the Mo chalcogenide-halides, they grow in the form of ordered strands, in which the individual strands are linked by very weak Van der Waals forces. Individual molecules can be manipulated, ordered and their length can be controlled with atomic force microscope tips.

Conduction of Electrons

Molecular wires conduct electricity. They typically have non-linear current-voltage characteristics, and do not behave as simple ohmic conductors. The conductance follows typical power law behavior as a function of temperature or electric field, whichever is the greater, arising from their strong one-dimensional character. Numerous theoretical ideas have been used in an attempt to understand the conductivity of one-dimensional systems, where strong interactions between electrons lead to departures from normal metallic (Fermi liquid) behavior. Important concepts are those introduced by Tomonaga, Luttinger and Wigner. Effects caused by classical Coulomb repulsion (called Coulomb blockade), interactions with vibrational degrees of freedom (called phonons) and Quantum Decoherence have also been found to be important in determining the properties of molecular wires.

Synthesis

Various methods have been developed for the synthesis of different types of molecular wires (e.g. organic molecular wires and inorganic molecular wires). The basic principle is to build repetitive molecular structures that are connected together to conduct electricity. Organic molecular wires are usually synthesized via transition metal-mediated cross-coupling reactions. Inorganic molecular wires are grown at conditions that allow formation of one-dimensional inorganic structures.

Organic Molecular Wires

Organic molecular wires usually consist aromatic rings connected by ethylene group or acetylene groups. Transition metal-mediated cross-coupling reactions are used to connect simple building blocks together in a convergent fashion to build organic molecular wires. For example, a simple oligo(phenylene ethylnylene) type molecular wire (B) was synthesized starting from readily available 1-bromo-4-iodobenzene (A). The final product was obtained through several steps of Sonogashira coupling reactions.

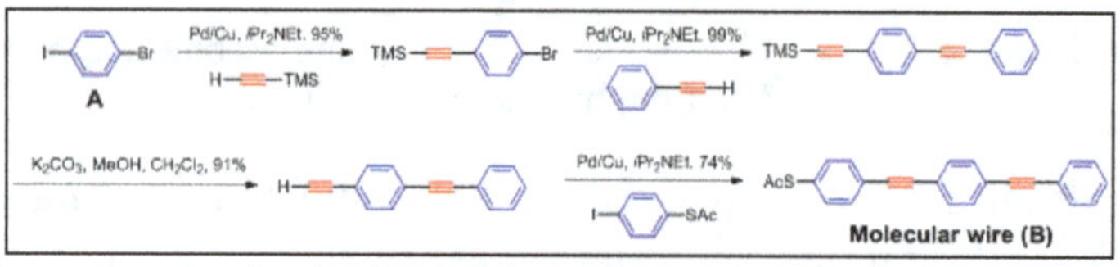

Synthesis of a simple organic molecular wire.

Other organic molecular wires include carbon nanotubes and DNA. Carbon nanotubes can be synthesized via various nano-technological approaches. DNA can be prepared by either step-wise DNA synthesis on solid-phase or by DNA-polymerase-catalyzed replication inside cells.

Inorganic Molecular Wires

Inorganic molecular wires consist of one-dimensional cluster structures that are usually held together by ionic bonds. These materials can be synthesized at high temperature. For example, a typical synthesis of Mo6S9-xIx was performed in sealed and vacuumed quartz ampoule. The commercially available molybdenum sheet metal, sulfur powder, and iodine were mixed at the desired ratio. The ampoule was slowly heated to a temperature of 1343 K, the sample was kept at that temperature for 72 hours and was then slowly cooled down to obtain the molecular wires.

Nanowires in Molecular Electronics

To be of use for connecting molecules, MWs need to self-assemble following well-defined routes and form reliable electrical contacts between them. To reproducibly self-assemble a complex circuit based on single molecules. Ideally, they would connect to diverse materials, such as gold metal surfaces (for connections to outside world), biomolecules (for nanosensors, nanoelectrodes, molecular switches) and most importantly, they must allow branching. The connectors should also be available of pre-determined diameter and length. They should also have covalent bonding to ensure reproducible transport and contact properties.

DNA-like molecules have specific molecular-scale recognition and can be used in molecular scaffold fabrication. Complex shapes have been demonstrated, but unfortunately metal coated DNA which is electrically conducting is too thick to connect to individual molecules. Thinner coated DNA lacks electronic connectivity and is unsuited for connecting molecular electronics components.

Some varieties of carbon nanotubes (CNTs) are conducting, and connectivity at their ends can be achieved by attachment of connecting groups. Unfortunately manufacturing CNTs with pre-determined properties is impossible at present, and the functionalized ends are typically not conducting, limiting their usefulness as molecular connectors. Individual CNTs can be soldered in an electron microscope, but the contact is not covalent and cannot be self-assembled.

Possible routes for the construction of larger functional circuits using $Mo_6S_{9-x}I_x$ MWs have been demonstrated, either via gold nanoparticles as linkers, or by direct connection to thiolated molecules. The two approaches may lead to different possible applications. The use of GNPs offers the possibility of branching and construction of larger circuits.

Molecular wires can be incorporated into polymers, enhancing their mechanical and/or conducting properties. The enhancement of these properties relies on uniform dispersion of the wires into the host polymer. MoSI wires have been made in such composites, relying on their superior solubility within the polymer host compared to other nanowires or nanotubes. Bundles of wires can be used to enhance tribological properties of polymers, with applications in actuators and potentiometers. It has been recently proposed that twisted nanowires could work as electromechanical nanodevices (or torsion nanobalances) to measure forces and torques at nanoscale with great precision.

MOLECULAR SCALE ELECTRONICS

Molecular scale electronics, also called single-molecule electronics, is a branch of nanotechnology that uses single molecules, or nanoscale collections of single molecules, as electronic components. Because single molecules constitute the smallest stable structures imaginable, this miniaturization is the ultimate goal for shrinking electrical circuits.

The field is often termed simply as "molecular electronics", but this term is also used to refer to the distantly related field of conductive polymers and organic electronics, which uses the properties of molecules to affect the bulk properties of a material. A nomenclature distinction has been suggested so that molecular materials for electronics refers to this latter field of bulk applications, while molecular scale electronics refers to the nanoscale single-molecule applications treated here.

Conventional electronics have traditionally been made from bulk materials. Ever since their invention in 1958, the performance and complexity of integrated circuits has undergone exponential growth, a trend named Moore's law, as feature sizes of the embedded components have shrunk accordingly. As the structures shrink, the sensitivity to deviations increases. In a few technology generations, when the minimum feature sizes reaches 13 nm, the composition of the devices must be controlled to a precision of a few atoms for the devices to work. With bulk methods growing increasingly demanding and costly as they near inherent limits, the idea was born that the components could instead be built up atom by atom in a chemistry lab (bottom up) versus carving them out of bulk material (top down). This is the idea behind molecular electronics, with the ultimate miniaturization being components contained in single molecules.

In single-molecule electronics, the bulk material is replaced by single molecules. Instead of forming structures by removing or applying material after a pattern scaffold, the atoms are put together in a chemistry lab. In this way, billions of billions of copies are made simultaneously (typically more than 10^{20} molecules are made at once) while the composition of molecules are controlled down to the last atom. The molecules used have properties that resemble traditional electronic components such as a wire, transistor or rectifier.

Single-molecule electronics is an emerging field, and entire electronic circuits consisting exclusively of molecular sized compounds are still very far from being realized. However, the unceasing demand for more computing power, along with the inherent limits of lithographic methods as of 2016, make the transition seem unavoidable. Currently, the focus is on discovering molecules with interesting properties and on finding ways to obtain reliable and reproducible contacts between the molecular components and the bulk material of the electrodes.

Theoretical Basis

Molecular electronics operates in the quantum realm of distances less than 100 nanometers. The miniaturization down to single molecules brings the scale down to a regime where quantum mechanics effects are important. In conventional electronic components, electrons can be filled in or drawn out more or less like a continuous flow of electric charge. In contrast, in molecular electronics the transfer of one electron alters the system significantly. For example, when an electron has been transferred from a source electrode to a molecule, the molecule gets charged up, which makes it far harder for the next electron to transfer. The significant amount of energy due to charging must be accounted for when making calculations about the electronic properties of the setup, and is highly sensitive to distances to conducting surfaces nearby.

The theory of single-molecule devices is especially interesting since the system under consideration is an open quantum system in nonequilibrium (driven by voltage). In the low bias voltage regime, the nonequilibrium nature of the molecular junction can be ignored, and the current-voltage traits of the device can be calculated using the equilibrium electronic structure of the system. However, in stronger bias regimes a more sophisticated treatment is required, as there is no longer a variational principle. In the elastic tunneling case (where the passing electron does not exchange energy with the system), the formalism of Rolf Landauer can be used to calculate the transmission through the system as a function of bias voltage, and hence the current. In inelastic tunneling, an elegant formalism based on the non-equilibrium Green's functions of Leo Kadanoff and Gordon Baym, and independently by Leonid Keldysh was advanced by Ned Wingreen and Yigal Meir. This Meir-Wingreen formulation has been used to great success in the molecular electronics community to examine the more difficult and interesting cases where the transient electron exchanges energy with the molecular system (for example through electron-phonon coupling or electronic excitations).

Further, connecting single molecules reliably to a larger scale circuit has proven a great challenge, and constitutes a significant hindrance to commercialization.

Examples:

Common for molecules used in molecular electronics is that the structures contain many alternating double and single bonds. This is done because such patterns delocalize the molecular orbitals, making it possible for electrons to move freely over the conjugated area.

Wires

The sole purpose of molecular wires is to electrically connect different parts of a molecular electrical circuit. As the assembly of these and their connection to a macroscopic circuit is still not mastered, the focus of research in single-molecule electronics is primarily on the functionalized molecules: molecular wires are characterized by containing no functional groups and are hence composed of plain repetitions of a conjugated building block. Among these are the carbon nanotubes that are quite large compared to the other suggestions but have shown very promising electrical properties.

The main problem with the molecular wires is to obtain good electrical contact with the electrodes so that electrons can move freely in and out of the wire.

Transistors

Single-molecule transistors are fundamentally different from the ones known from bulk electronics. The gate in a conventional (field-effect) transistor determines the conductance between the source and drain electrode by controlling the density of charge carriers between them, whereas the gate in a single-molecule transistor controls the possibility of a single electron to jump on and off the molecule by modifying the energy of the molecular orbitals. One of the effects of this difference is that the single-molecule transistor is almost binary: It is either on or off. This opposes its bulk counterparts, which have quadratic responses to gate voltage.

It is the quantization of charge into electrons that is responsible for the markedly different behavior compared to bulk electronics. Because of the size of a single molecule, the charging due to a single electron is significant and provides means to turn a transistor on or off. For this to work, the electronic orbitals on the transistor molecule cannot be too well integrated with the orbitals on the electrodes. If they are, an electron cannot be said to be located on the molecule or the electrodes and the molecule will function as a wire.

A popular group of molecules, that can work as the semiconducting channel material in a molecular transistor, is the oligopolyphenylenevinylenes (OPVs) that works by the Coulomb blockade mechanism when placed between the source and drain electrode in an appropriate way. Fullerenes work by the same mechanism and have also been commonly used.

Semiconducting carbon nanotubes have also been demonstrated to work as channel material but although molecular, these molecules are sufficiently large to behave almost as bulk semiconductors.

The size of the molecules, and the low temperature of the measurements being conducted, makes the quantum mechanical states well defined. Thus, it is being researched if the quantum mechanical properties can be used for more advanced purposes than simple transistors (e.g., spintronics).

Physicists at the University of Arizona, in collaboration with chemists from the University of Madrid, have designed a single-molecule transistor using a ring-shaped molecule similar to benzene. Physicists at Canada's National Institute for Nanotechnology have designed a single-molecule transistor using styrene. Both groups expect (the designs were experimentally unverified as of June 2005) their respective devices to function at room temperature, and to be controlled by a single electron.

Rectifiers (Diodes)

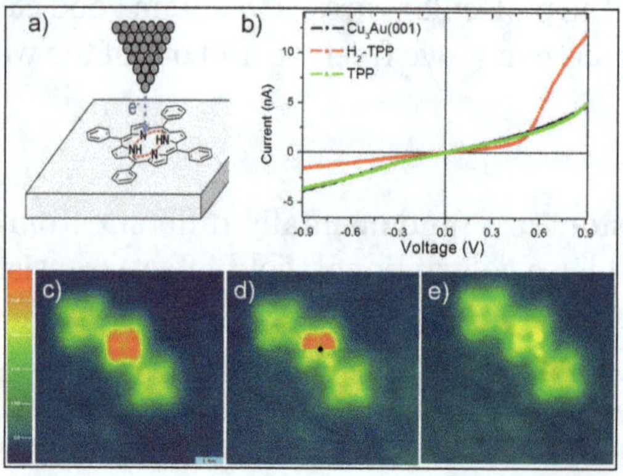

Hydrogen can be removed from individual tetraphenylporphyrin (H_2TPP) molecules by applying excess voltage to the tip of a scanning tunneling microscope (STAM, a); this removal alters the current-voltage (I-V) curves of TPP molecules, measured using the same STM tip, from diode-like (red curve in b) to resistor-like (green curve). Image (c) shows a row of TPP, H_2TPP and TPP molecules. While scanning image (d), excess voltage was applied to H_2TPP at the black dot, which instantly removed hydrogen, as shown in the bottom part of (d) and in the re-scan image (e). Such manipulations can be used in single-molecule electronics.

Molecular rectifiers are mimics of their bulk counterparts and have an asymmetric construction so that the molecule can accept electrons in one end but not the other. The molecules have an electron donor (D) in one end and an electron acceptor (A) in the other. This way, the unstable state $D^+ - A^-$ will be more readily made than $D^- - A^+$. The result is that an electric current can be drawn through the molecule if the electrons are added through the acceptor end, but less easily if the reverse is attempted.

Methods

One of the biggest problems with measuring on single molecules is to establish reproducible electrical contact with only one molecule and doing so without shortcutting the electrodes. Because the current photolithographic technology is unable to produce electrode gaps small enough to contact both ends of the molecules tested (on the order of nanometers), alternative strategies are applied.

Molecular Gaps

One way to produce electrodes with a molecular sized gap between them is break junctions, in which a thin electrode is stretched until it breaks. Another is electromigration. Here a current is led through a thin wire until it melts and the atoms migrate to produce the gap. Further, the reach of conventional photolithography can be enhanced by chemically etching or depositing metal on the electrodes.

Probably the easiest way to conduct measurements on several molecules is to use the tip of a scanning tunneling microscope (STM) to contact molecules adhered at the other end to a metal substrate.

Anchoring

A popular way to anchor molecules to the electrodes is to make use of sulfur's high chemical affinity to gold. In these setups, the molecules are synthesized so that sulfur atoms are placed strategically to function as crocodile clips connecting the molecules to the gold electrodes. Though useful, the anchoring is non-specific and thus anchors the molecules randomly to all gold surfaces. Further, the contact resistance is highly dependent on the precise atomic geometry around the site of anchoring and thereby inherently compromises the reproducibility of the connection.

To circumvent the latter issue, experiments has shown that fullerenes could be a good candidate for use instead of sulfur because of the large conjugated π-system that can electrically contact many more atoms at once than one atom of sulfur.

Fullerene Nanoelectronics

In polymers, classical organic molecules are composed of both carbon and hydrogen (and sometimes additional compounds such as nitrogen, chlorine or sulphur). They are obtained from petrol and can often be synthesized in large amounts. Most of these molecules are insulating when their length exceeds a few nanometers. However, naturally occurring carbon is conducting, especially graphite recovered from coal or encountered otherwise. From a theoretical viewpoint, graphite is a semi-metal, a category in between metals and semi-conductors. It has a layered structure, each sheet being one atom thick. Between each sheet, the interactions are weak enough to allow an easy manual cleavage.

Tailoring the graphite sheet to obtain well defined nanometer-sized objects remains a challenge. However, by the close of the twentieth century, chemists were exploring methods to fabricate extremely small graphitic objects that could be considered single molecules. After studying the interstellar conditions under which carbon is known to form clusters, Richard Smalley's group (Rice University, Texas) set up an experiment in which graphite was vaporized via laser irradiation. Mass spectrometry revealed that clusters containing specific *magic numbers* of atoms were stable, especially those clusters of 60 atoms. Harry Kroto, an English chemist who assisted in the experiment, suggested a possible geometry for these clusters – atoms covalently bound with the exact symmetry of a soccer ball. Coined buckminsterfullerenes, buckyballs, or C_{60}, the clusters retained some properties of graphite, such as conductivity. These objects were rapidly envisioned as possible building blocks for molecular electronics.

Problems

Artifacts

When trying to measure electronic traits of molecules, artificial phenomena can occur that can be hard to distinguish from truly molecular behavior. Before they were discovered, these artifacts have mistakenly been published as being features pertaining to the molecules in question.

Applying a voltage drop on the order of volts across a nanometer sized junction results in a very strong electrical field. The field can cause metal atoms to migrate and eventually close the gap by a thin filament, which can be broken again when carrying a current. The two levels of conductance imitate molecular switching between a conductive and an isolating state of a molecule.

Another encountered artifact is when the electrodes undergo chemical reactions due to the high field strength in the gap. When the voltage bias is reversed, the reaction will cause hysteresis in the measurements that can be interpreted as being of molecular origin.

A metallic grain between the electrodes can act as a single electron transistor by the mechanism described above, thus resembling the traits of a molecular transistor. This artifact is especially common with nanogaps produced by the electromigration method.

Commercialization

One of the biggest hindrances for single-molecule electronics to be commercially exploited is the lack of methods to connect a molecular sized circuit to bulk electrodes in a way that gives reproducible results. At the current state, the difficulty of connecting single molecules vastly outweighs any possible performance increase that could be gained from such shrinkage. The difficulties grow worse if the molecules are to have a certain spatial orientation and/or have multiple poles to connect.

Also problematic is that some measurements on single molecules are carried out in cryogenic temperatures (near absolute zero), which is very energy consuming. This is done to reduce signal noise enough to measure the faint currents of single molecules.

NANOCIRCUITRY AND NANOCIRCUITS

Nanocircuits are electrical circuits operating on the nanometer scale. This is well into the quantum realm, where quantum mechanical effects become very important. One nanometer is equal to 10^{-9} meters or a row of 10 hydrogen atoms. With such progressively smaller circuits, more can be fitted on a computer chip. This allows faster and more complex functions using less power. Nanocircuits are composed of three different fundamental components. These are transistors, interconnections, and architecture, all fabricated on the nanometer scale.

Various Approaches to Nanocircuitry

A variety of proposals have been made to implement nanocircuitry in different forms. These include Nanowires, Single-Electron Transistors, Quantum dot cellular automata, and Nanoscale Crossbar Latches. However, likely nearer-term approaches will involve incorporation of nanomaterials to improve MOSFETs (metal-oxide-semiconductor field-effect transistors). These currently form the basis of most analog and digital circuit designs, the scaling of which drives Moore's Law. A review article covering the MOSFET design and its future was published in 2004 comparing different geometries of MOSFETs under scale reduction and noted that circular cross-section vertical channel FETs are optimal for scale reduction. This configuration is capable of being implemented with a high density using vertical semiconductor cylindrical channels with nanoscale diameters and Infineon Technologies and Samsung have begun research and development in this direction resulting in some basic patents using nanowires and carbon nanotubes in MOSFET designs. In an alternative approach, Nanosys uses solution based deposition and alignment processes to pattern pre-fabricated arrays of nanowires on a substrate to serve as a lateral channel of an FET. While not capable of the same scalability as single nanowire FETs, the use of pre-fabricated multiple nanowires for the channel increases reliability and reduces production costs since large volume printing processes may be used to deposit the nanowires at a lower temperature than conventional fabrication procedures. In addition, due to the lower temperature deposition a wider variety of materials such as polymers may be used as the carrier substrate for the transistors opening the door to flexible electronic applications such as electronic paper, bendable flat panel displays, and wide area solar cells.

Production Methods

One of the most fundamental concepts to understanding nanocircuits is the formulation of Moore's Law. This concept arose when Intel co-founder Gordon Moore became interested in the cost of transistors and trying to fit more onto one chip. It relates that the number of transistors that can be fabricated on a silicon integrated circuit—and therefore the computing abilities of such a circuit—is doubling every 18 to 24 months. The more transistors one can fit on a circuit, the more computational abilities the computer will have. This is why scientists and engineers are working together to produce these nanocircuits so increasingly more and more transistors will be able to fit onto a chip. Despite how good this may sound, there are many problems that arise when so many transistors are packed together. With circuits being so tiny, they tend to have more problems than larger circuits, more particularly heat - the amount of power applied over a smaller surface area makes heat dissipation difficult, this excess heat will cause errors and can destroy the chip. Nanoscale circuits are more sensitive to temperature changes, cosmic rays and electromagnetic interference than today's circuits. As more transistors are packed onto a chip, phenomena such as stray signals on the chip, the need to dissipate the heat from so many closely packed devices, tunneling across insulation barriers due to the small scale, and fabrication difficulties will halt or severely slow progress. Many believe the market for nanocircuits will reach equilibrium around 2015. At this time they believe the cost of a fabrication facility may be as much as $200 billion. There will be a time when the cost of making circuits even smaller will be too much, and the speed of computers will reach a maximum. For this reason, many scientists believe that Moore's Law will not hold forever and will soon reach a peak, since Moore's law is largely predicated on computational gains caused by improvements in micro-lithographic etching technologies.

In producing these nanocircuits, there are many aspects involved. The first part of their organization begins with transistors. As of right now, most electronics are using silicon-based transistors. Transistors are an integral part of circuits as they control the flow of electricity and transform weak electrical signals to strong ones. They also control electric current as they can turn it on off, or even amplify signals. Circuits now use silicon as a transistor because it can easily be switched between conducting and nonconducting states. However, in nanoelectronics, transistors might be organic molecules or nanoscale inorganic structures. Semiconductors, which are part of transistors, are also being made of organic molecules in the nano state.

The second aspect of nanocircuit organization is interconnection. This involves logical and mathematical operations and the wires linking the transistors together that make this possible. In nanocircuits, nanotubes and other wires as narrow as one nanometer are used to link transistors together. Nanowires have been made from carbon nanotubes for a few years. Until a few years ago, transistors and nanowires were put together to produce the circuit. However, scientists have been able to produce a nanowire with transistors in it. In 2004, Harvard University nanotech pioneer Charles Lieber and his team have made a nanowire—10,000 times thinner than a sheet of paper—that contains a string of transistors. Essentially, transistors and nanowires are already pre-wired so as to eliminate the difficult task of trying to connect transistors together with nanowires.

The last part of nanocircuit organization is architecture. This has been explained as the overall way the transistors are interconnected, so that the circuit can plug into a computer or other system and operate independently of the lower-level details. With nanocircuits being so small, they are destined for error and defects. Scientists have devised a way to get around this. Their architecture combines circuits that have redundant logic gates and interconnections with the ability to reconfigure structures at several levels on a chip. The redundancy lets the circuit identify problems and reconfigure itself so the circuit can avoid more problems. It also allows for errors within the logic gate and still have it work properly without giving a wrong result.

Experimental Breakthroughs and Potential Applications

In 1960, Egyptian engineer Mohamed Atalla and Korean engineer Dawon Kahng at Bell Labs fabricated the first MOSFET (metal-oxide-semiconductor field-effect transistor) with a gate oxide thickness of 100 nm, along with a gate length of 20 μm. In 1962, Atalla and Kahng fabricated a nanolayer-base metal–semiconductor junction (M–S junction) transistor that used gold (Au) thin films with a thickness of 10 nm.

In 1987, Iranian engineer Bijan Davari led an IBM research team that demonstrated the first MOSFET with a 10 nm gate oxide thickness, using tungsten-gate technology. Multi-gate MOSFETs enabled scaling below 20 nm gate length, starting with the FinFET (fin field-effect transistor), a three-dimensional, non-planar, double-gate MOSFET. The FinFET originates from the research of Digh Hisamoto at Hitachi Central Research Laboratory in 1989. At UC Berkeley, FinFET devices were fabricated by a group consisting of Hisamoto along with TSMC's Chenming Hu and other international researchers including Tsu-Jae King Liu, Jeffrey Bokor, Hideki Takeuchi, K. Asano,

Jakub Kedziersk, Xuejue Huang, Leland Chang, Nick Lindert, Shibly Ahmed and Cyrus Tabery. The team fabricated FinFET devices down to a 17 nm process in 1998, and then 15 nm in 2001. In 2002, a team including Yu, Chang, Ahmed, Hu, Liu, Bokor and Tabery fabricated a 10 nm FinFET device.

In 2005, Indian physicists Prabhakar Bandaru and Apparao M. Rao at UC San Diego developed the world's smallest transistor based to be made entirely from carbon nanotubes. It was intended to be used for nanocircuits. Nanotubes are rolled up sheets of carbon atoms and are more than a thousand times thinner than human hair. In 2006, a team of Korean researchers from the Korea Advanced Institute of Science and Technology (KAIST) and the National Nano Fab Center developed a 3 nm MOSFET, the world's smallest nanoelectronic device, based on gate-all-around (GAA) FinFET technology.

Normally, circuits use silicon-based transistors, but carbon nanotubes are intended to replace those. The transistor has two different branches that meet at a single point, hence giving it a Y shape. Current can flow throughout both branches and is controlled by a third branch that turns the voltage on or off. This new breakthrough can now allow for nanocircuits to hold completely to their name as they can be made entirely from nanotubes. Before this discovery, logic circuits used nanotubes, but needed metal gates to be able to control the flow of electric current.

Arguably the biggest potential application of nanocircuits deals with computers and electronics. Scientists and engineers are always looking to make computers faster. Some think in the nearer term, we could see hybrids of micro- and nano-: silicon with a nano core—perhaps a high-density computer memory that retains its contents forever. Unlike conventional circuit design, which proceeds from blueprint to photographic pattern to chip, nanocircuit design will probably begin with the chip—a haphazard jumble of as many as 1024 components and wires, not all of which will even work—and gradually sculpt it into a useful device. Instead of taking the traditional top-down approach, the bottom-up approach will probably soon have to be adopted because of the sheer size of these nanocircuits. Not everything in the circuit will probably work because at the nano level, nanocircuits will be more defective and faulty because of their compactness. Scientists and engineers have created all of the essential components of nanocircuits such as transistors, logic gates and diodes. They have all been constructed from organic molecules, carbon nanotubes and nanowire semiconductors. The only thing left to do is find a way to eliminate the errors that come with such a small device and nanocircuits will become a way of all electronics. However, eventually there will be a limit as to how small nanocircuits can become and computers and electronics will reach their equilibrium speeds.

NANOWIRE

A nanowire is a nanostructure, with the diameter of the order of a nanometer (10^{-9}

meters). It can also be defined as the ratio of the length to width being greater than 1000. Alternatively, nanowires can be defined as structures that have a thickness or diameter constrained to tens of nanometers or less and an unconstrained length. At these scales, quantum mechanical effects are important—which coined the term "quantum wires". Many different types of nanowires exist, including superconducting (e.g., YBCO), metallic (e.g. Ni, Pt, Au), semiconducting (e.g. silicon nanowires (SiNWs), InP, GaN) and insulating (e.g. SiO_2, TiO_2). Molecular nanowires are composed of repeating molecular units either organic (e.g. DNA) or inorganic (e.g. $Mo_6S_{9-x}I_x$).

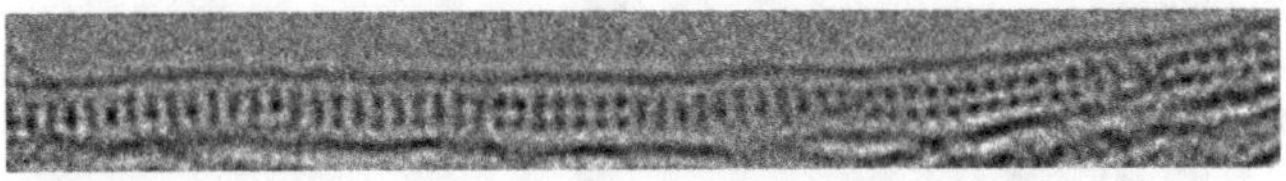

Crystalline 2×2-atom tin selenide nanowire grown inside a single-wall carbon nanotube (tube diameter ~1 nm).

Typical nanowires exhibit aspect ratios (length-to-width ratio) of 1000 or more. As such they are often referred to as one-dimensional (1-D) materials. Nanowires have many interesting properties that are not seen in bulk or 3-D (three-dimensional) materials. This is because electrons in nanowires are quantum confined laterally and thus occupy energy levels that are different from the traditional continuum of energy levels or bands found in bulk materials.

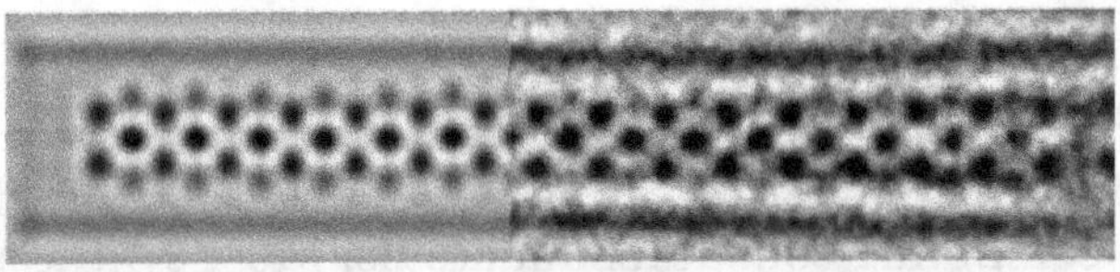

A noise-filtered HRTEM image of a HgTe extreme nanowire embedded down the central pore of a SWCNT. The image is also accompanied by a simulation of the crystal structure.

Peculiar features of this quantum confinement exhibited by certain nanowires manifest themselves in discrete values of the electrical conductance. Such discrete values arise from a quantum mechanical restraint on the number of electrons that can travel through the wire at the nanometer scale. These discrete values are often referred to as the quantum of conductance and are integer multiples of:

$$\frac{2e^2}{h} \simeq 77.41\,\mu S.$$

They are inverse of the well-known resistance unit h/e^2, which is roughly equal to 25812.8 ohms, and referred to as the von Klitzing constant R_K (after Klaus von Klitzing, the discoverer of exact quantization). Since 1990, a fixed conventional value R_{K-90} is accepted.

Examples of nanowires include inorganic molecular nanowires ($Mo_6S_{9-x}I_x$, $Li_2Mo_6Se_6$), which can have a diameter of 0.9 nm and be hundreds of micrometers long. Other important examples are based on semiconductors such as InP, Si, GaN, etc., dielectrics (e.g. SiO_2,TiO_2), or metals (e.g. Ni, Pt).

There are many applications where nanowires may become important in electronic, opto-electronic and nanoelectromechanical devices, as additives in advanced composites, for metallic interconnects in nanoscale quantum devices, as field-emitters and as leads for biomolecular nanosensors.

Synthesis

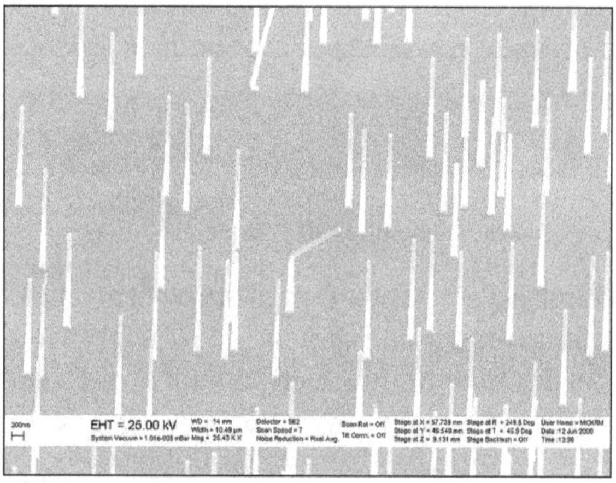

An SEM image of epitaxial nanowire heterostructures grown from catalytic gold nanoparticles.

There are two basic approaches to synthesizing nanowires: top-down and bottom-up. A top-down approach reduces a large piece of material to small pieces, by various means such as lithography, milling or thermal oxidation. A bottom-up approach synthesizes the nanowire by combining constituent adatoms. Most synthesis techniques use a bottom-up approach. Initial synthesis via either method may often be followed by a nanowire thermal treatment step, often involving a form of self-limiting oxidation, to fine tune the size and aspect ratio of the structures.

Nanowire production uses several common laboratory techniques, including suspension, electrochemical deposition, vapor deposition, and VLS growth. Ion track technology enables growing homogeneous and segmented nanowires down to 8 nm diameter. As nanowire oxidation rate is controlled by diameter, thermal oxidation steps are often applied to tune their morphology.

Suspension

A suspended nanowire is a wire produced in a high-vacuum chamber held at the longitudinal extremities. Suspended nanowires can be produced by:

- The chemical etching of a larger wire.
- The bombardment of a larger wire, typically with highly energetic ions.
- Indenting the tip of a STM in the surface of a metal near its melting point, and then retracting it.

VLS Growth

A common technique for creating a nanowire is vapor-liquid-solid method (VLS), which was first reported by Wagner and Ellis in 1964 for silicon whiskers with diameters ranging from hundreds of nm to hundreds of μm. This process can produce high-quality crystalline nanowires of many semiconductor materials, for example, VLS–grown single crystalline silicon nanowires (SiNWs) with smooth surfaces could have excellent properties, such as ultra-large elasticity. This method uses a source material from either laser ablated particles or a feed gas such as silane.

VLS synthesis requires a catalyst. For nanowires, the best catalysts are liquid metal (such as gold) nanoclusters, which can either be self-assembled from a thin film by dewetting, or purchased in colloidal form and deposited on a substrate.

The source enters these nanoclusters and begins to saturate them. On reaching supersaturation, the source solidifies and grows outward from the nanocluster. Simply turning off the source can adjust the final length of the nanowire. Switching sources while still in the growth phase can create compound nanowires with super-lattices of alternating materials.

A single-step vapour phase reaction at elevated temperature synthesises inorganic nanowires such as $Mo_6S_{9-x}I_x$. From another point of view, such nanowires are cluster polymers.

VSS Growth Similar to VLS synthesis, VSS (vapor-solid-solid) synthesis of nanowires (NWs) proceeds through thermolytic decomposition of a silicon precursor (typically phenylsilane). Unlike VLS, the catalytic seed remains in solid state when subjected to high temperature annealing of the substrate. This such type of synthesis is widely used to synthesise metal silicide/germanide nanowires through VSS alloying between a copper substrate and a silicon/germanium precursor.

Solution-phase Synthesis

Solution-phase synthesis refers to techniques that grow nanowires in solution. They can produce nanowires of many types of materials. Solution-phase synthesis has the advantage that it can produce very large quantities, compared to other methods. In one technique, the polyol synthesis, ethylene glycol is both solvent and reducing agent. This technique is particularly versatile at producing nanowires of gold, lead, platinum, and silver.

The supercritical fluid-liquid-solid growth method can be used to synthesize semiconductor nanowires, e.g., Si and Ge. By using metal nanocrystals as seeds, Si and Ge organometallic precursors are fed into a reactor filled with a supercritical organic solvent, such as toluene. Thermolysis results in degradation of the precursor, allowing release of Si or Ge, and dissolution into the metal nanocrystals. As more of the semiconductor solute is added from the supercritical phase (due to a concentration gradient), a solid crystallite precipitates, and a nanowire grows uniaxially from the nanocrystal seed.

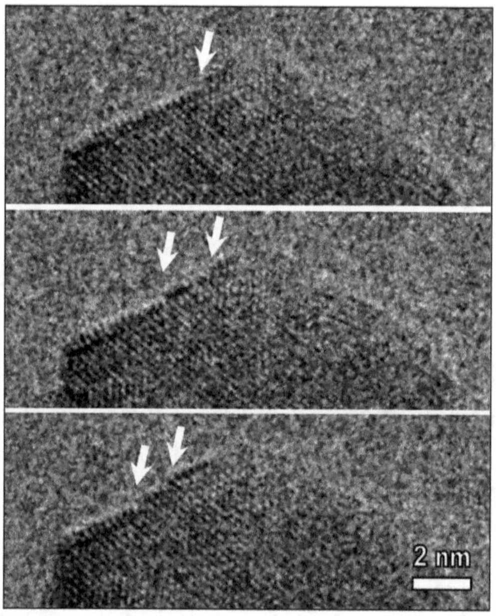

In situ observation of CuO nanowire growth.

Non-catalytic Growth

Nanowires can be also grown without the help of catalysts, which gives an advantage of pure nanowires and minimizes the number of technological steps. The simplest methods to obtain metal oxide nanowires use ordinary heating of the metals, e.g. metal wire heated with battery, by Joule heating in air can be easily done at home. The vast majority of nanowire-formation mechanisms are explained through the use of catalytic nanoparticles, which drive the nanowire growth and are either added intentionally or generated during the growth. However the mechanisms for catalyst-free growth of nanowires (or whiskers) were known from 1950s. Spontaneous nanowire formation by non-catalytic methods were explained by the dislocation present in specific directions or the growth anisotropy of various crystal faces. More recently, after microscopy advancement, the nanowire growth driven by screw dislocations or twin boundaries were demonstrated. The picture on the right shows a single atomic layer growth on the tip of CuO nanowire, observed by in situ TEM microscopy during the non-catalytic synthesis of nanowire.

DNA-templated Metallic Nanowire Synthesis

An emerging field is to use DNA strands as scaffolds for metallic nanowire synthesis. This method is investigated both for the synthesis of metallic nanowires in electronic components and for biosensing applications, in which they allow the transduction of a

DNA strand into a metallic nanowire that can be electrically detected. Typically, ssDNA strands are stretched, whereafter they are decorated with metallic nanoparticles that have been functionalised with short complementary ssDNA strands.

Physics

Conductivity of Nanowires

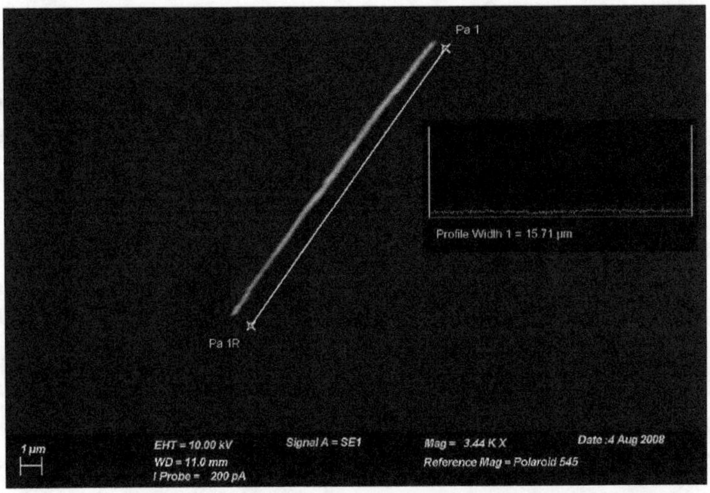

An SEM image of a 15 micrometer nickel wire.

Several physical reasons predict that the conductivity of a nanowire will be much less than that of the corresponding bulk material. First, there is scattering from the wire boundaries, whose effect will be very significant whenever the wire width is below the free electron mean free path of the bulk material. In copper, for example, the mean free path is 40 nm. Copper nanowires less than 40 nm wide will shorten the mean free path to the wire width.

Nanowires also show other peculiar electrical properties due to their size. Unlike single wall carbon nanotubes, whose motion of electrons can fall under the regime of ballistic transport (meaning the electrons can travel freely from one electrode to the other), nanowire conductivity is strongly influenced by edge effects. The edge effects come from atoms that lay at the nanowire surface and are not fully bonded to neighboring atoms like the atoms within the bulk of the nanowire. The unbonded atoms are often a source of defects within the nanowire, and may cause the nanowire to conduct electricity more poorly than the bulk material. As a nanowire shrinks in size, the surface atoms become more numerous compared to the atoms within the nanowire, and edge effects become more important.

Furthermore, the conductivity can undergo a quantization in energy: i.e. the energy of the electrons going through a nanowire can assume only discrete values, which are multiples of the conductance quantum $G = 2e^2/h$ (where e is the charge of the electron and h is the Planck constant. See also the Quantum Hall effect).

The conductivity is hence described as the sum of the transport by separate *channels* of different quantized energy levels. The thinner the wire is, the smaller the number of channels available to the transport of electrons.

This quantization has been demonstrated by measuring the conductivity of a nanowire suspended between two electrodes while pulling it: as its diameter reduces, its conductivity decreases in a stepwise fashion and the plateaus correspond to multiples of G.

The quantization of conductivity is more pronounced in semiconductors like Si or GaAs than in metals, due to their lower electron density and lower effective mass. It can be observed in 25 nm wide silicon fins, and results in increased threshold voltage. In practical terms, this means that a MOSFET with such nanoscale silicon fins, when used in digital applications, will need a higher gate (control) voltage to switch the transistor on.

Welding Nanowires

To incorporate nanowire technology into industrial applications, researchers in 2008 developed a method of welding nanowires together: A sacrificial metal nanowire is placed adjacent to the ends of the pieces to be joined (using the manipulators of a scanning electron microscope); then an electric current is applied, which fuses the wire ends. The technique fuses wires as small as 10 nm.

For nanowires with diameters less than 10 nm, existing welding techniques, which require precise control of the heating mechanism and which may introduce the possibility of damage, will not be practical. Recently scientists discovered that single-crystalline ultrathin gold nanowires with diameters ~3–10 nm can be "cold-welded" together within seconds by mechanical contact alone, and under remarkably low applied pressures (unlike macro- and micro-scale cold welding process). High-resolution transmission electron microscopy and in situ measurements reveal that the welds are nearly perfect, with the same crystal orientation, strength and electrical conductivity as the rest of the nanowire. The high quality of the welds is attributed to the nanoscale sample dimensions, oriented-attachment mechanisms and mechanically assisted fast surface diffusion. Nanowire welds were also demonstrated between gold and silver, and silver nanowires (with diameters ~5–15 nm) at near room temperature, indicating that this technique may be generally applicable for ultrathin metallic nanowires. Combined with other nano- and microfabrication technologies, cold welding is anticipated to have potential applications in the future bottom-up assembly of metallic one-dimensional nanostructures.

Mechanical Properties of Nanowires

Investigation of Mechanical Properties of Nanowires

The study of nanowire mechanics has boomed since the advent of the Atomic Force Microscope (AFM), and associated technologies which have enabled direct study of the response of the nanowire to an applied load. Specifically, a nanowire can be clamped from one end, and the free end displaced by an AFM tip. In this cantilever geometry, the height of the AFM is precisely known, and the force applied is precisely known. This allows for construction of a force vs. displacement curve, which can be converted to a stress vs. strain curve if the nanowire dimensions are known. From the stress-strain curve, the elastic constant known as the Young's Modulus can be derived, as well as the toughness, and degree of strain-hardening.

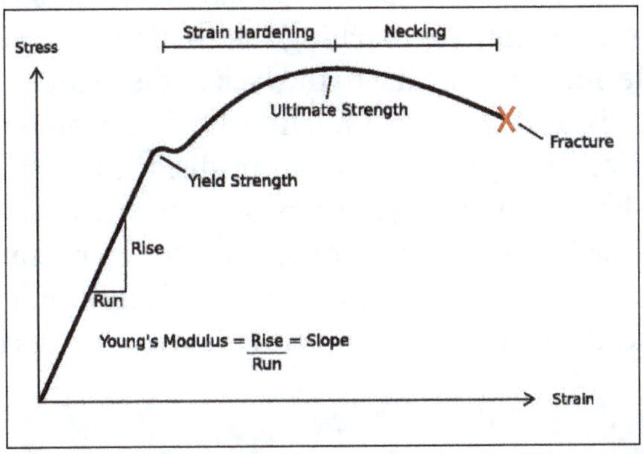

Stress-strain curve provides all the relevant mechanical properties including; tensile modulus, yield strength, ultimate tensile strength, and fracture strength.

Young's Modulus of Nanowires

The elastic component of the stress-strain curve described by the Young's Modulus, has been reported for nanowires, however the modulus depends very strongly on the microstructure. Thus a complete description of the modulus dependence on diameter is lacking. Analytically, continuum mechanics has been applied to estimate the dependence of modulus on diameter: $E = E_o[1+4(E_o / E_s - 1)(r_s / D - r_s^2 / D^2)]$ in tension, where E_o is the bulk modulus, r_s is the thickness of a shell layer in which the modulus is surface dependent and varies from the bulk, *Es* is the surface modulus, and *D* is the diameter. This equation implies that the modulus increases as the diameter decreases. However, various computational methods such as molecular dynamics have predicted that modulus should decrease as diameter decreases.

Experimentally, gold nanowires have been shown to have a Young's modulus which is effectively diameter independent. Similarly, nano-indentation was applied to study the modulus of silver nanowires, and again the modulus was found to be 88 GPa, very similar to the modulus of bulk Silver (85 GPa) These works demonstrated that the analytically determined modulus dependence seems to be suppressed in nanowire samples where the crystalline structure highly resembles that of the bulk system.

In contrast, Si solid nanowires have been studied, and shown to have a decreasing modulus with diameter The authors of that work report a Si modulus which is half that of the bulk value, and they suggest that the density of point defects, and or loss of chemical stoichiometry may account for this difference.

Yield Strength of Nanowires

The plastic component of the stress strain curve (or more accurately the onset of plasticity) is described by the yield strength. The strength of a material is increased by decreasing the number of defects in the solid, which occurs naturally in nanomaterials where the volume of the solid is reduced. As a nanowire is shrunk to a single line of atoms, the strength should theoretically increase all the way to the molecular tensile strength. Gold nanowires have been described as 'ultrahigh strength' due to the extreme increase in yield strength, approaching the theoretical value of E/10. This huge increase in yield is determined to be due to the lack of dislocations in the solid. Without dislocation motion, a 'dislocation-starvation' mechanism is in operation. The material can accordingly experience huge stresses before dislocation motion is possible, and then begins to strain-harden. For these reasons, nanowires (historically described as 'whiskers') have been used extensively in composites for increasing the overall strength of a material. Moreover, nanowires continue to be actively studied, with research aiming to translate enhanced mechanical properties to novel devices in the fields of MEMS or NEMS.

Applications

Electronic Devices

Nanowires can be used for MOSFETs (MOS field-effect transistors). MOS transistors are used widely as fundamental building elements in today's electronic circuits. As predicted by Moore's law, the dimension of MOS transistors is shrinking smaller and smaller into nanoscale. One of the key challenges of building future nanoscale MOS transistors is ensuring good gate control over the channel. Due to the high aspect ratio, if the gate dielectric is wrapped around the nanowire channel, we can get good control of channel electrostatic potential, thereby turning the transistor on and off efficiently.

Due to the unique one-dimensional structure with remarkable optical properties, the nanowire also opens new opportunities for realizing high efficiency photovoltaic devices. Compared with its bulk counterparts, the nanowire solar cells are less sensitive to impurities due to bulk recombination, and thus silicon wafers with lower purity can be used to achieve acceptable efficiency, leading to the a reduction on material consumption.

To create active electronic elements, the first key step was to chemically dope a semiconductor nanowire. This has already been done to individual nanowires to create p-type and n-type semiconductors.

The next step was to find a way to create a p–n junction, one of the simplest electronic devices. This was achieved in two ways. The first way was to physically cross a p-type wire over an n-type wire. The second method involved chemically doping a single wire with different dopants along the length. This method created a p-n junction with only one wire.

After p-n junctions were built with nanowires, the next logical step was to build logic gates. By connecting several p-n junctions together, researchers have been able to create the basis of all logic circuits: the AND, OR, and NOT gates have all been built from semiconductor nanowire crossings.

In August 2012, researchers reported constructing the first NAND gate from undoped silicon nanowires. This avoids the problem of how to achieve precision doping of complementary nanocircuits, which is unsolved. They were able to control the Schottky barrier to achieve low-resistance contacts by placing a silicide layer in the metal-silicon interface.

It is possible that semiconductor nanowire crossings will be important to the future of digital computing. Though there are other uses for nanowires beyond these, the only ones that actually take advantage of physics in the nanometer regime are electronic.

In addition, nanowires are also being studied for use as photon ballistic waveguides as interconnects in quantum dot/quantum effect well photon logic arrays. Photons travel inside the tube, electrons travel on the outside shell.

When two nanowires acting as photon waveguides cross each other the juncture acts as a quantum dot.

Conducting nanowires offer the possibility of connecting molecular-scale entities in a molecular computer. Dispersions of conducting nanowires in different polymers are being investigated for use as transparent electrodes for flexible flat-screen displays.

Because of their high Young's moduli, their use in mechanically enhancing composites is being investigated. Because nanowires appear in bundles, they may be used as tribological additives to improve friction characteristics and reliability of electronic transducers and actuators.

Because of their high aspect ratio, nanowires are also uniquely suited to dielectrophoretic manipulation, which offers a low-cost, bottom-up approach to integrating suspended dielectric metal oxide nanowires in electronic devices such as UV, water vapor, and ethanol sensors.

Nanowire Lasers

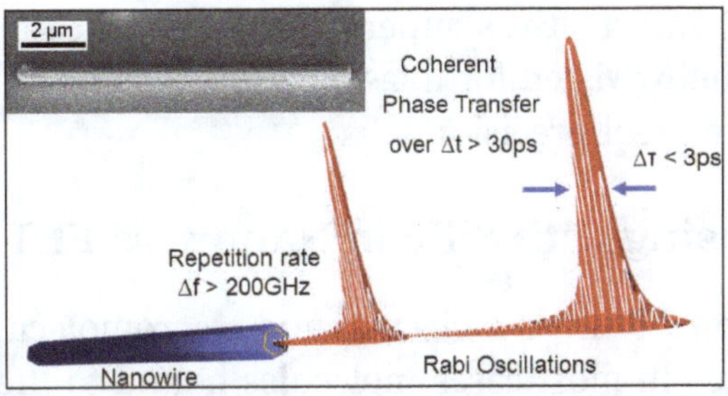

Nanowire lasers for ultrafast transmission of information in light pulses.

Nanowire lasers are nano-scaled lasers with potential as optical interconnects and optical data communication on chip. Nanowire lasers are built from III–V semiconductor heterostructures, the high refractive index allows for low optical loss in the nanowire core. Nanowire lasers are subwavelength lasers of only a few hundred nanometers. Nanowire lasers are Fabry–Perot resonator cavities defined by the end facets of the wire with high-reflectivity, recent developments have demonstrated repetition rates greater than 200 GHz offering possibilities for optical chip level communications.

Sensing of Proteins and Chemicals using Semiconductor Nanowires

In an analogous way to FET devices in which the modulation of conductance (flow of electrons/holes) in the semiconductor, between the input (source) and the output (drain) terminals, is controlled by electrostatic potential variation (gate-electrode) of the charge carriers in the device conduction channel, the methodology of a Bio/Chem-FET is based on the detection of the local change in charge density, or so-called "field effect", that characterizes the recognition event between a target molecule and the surface receptor.

This change in the surface potential influences the Chem-FET device exactly as a 'gate' voltage does, leading to a detectable and measurable change in the device conduction. When these devices are fabricated using semiconductor nanowires as the transistor element the binding of a chemical or biological species to the surface of the sensor can lead to the depletion or accumulation of charge carriers in the "bulk" of the nanometer diameter nanowire i.e. (small cross section available for conduction channels). Moreover, the wire, which serves as a tunable conducting channel, is in close contact with the sensing environment of the target, leading to a short response time, along with orders of magnitude increase in the sensitivity of the device as a result of the huge S/V ratio of the nanowires.

While several inorganic semiconducting materials such as Si, Ge, and metal oxides (e.g. In_2O_3, S_nO_2, Z_nO, etc.) have been used for the preparation of nanowires, Si is usually the material of choice when fabricating nanowire FET-based chemo/biosensors.

Several examples of the use of silicon nanowire(SiNW) sensing devices include the ultra sensitive, real-time sensing of biomarker proteins for cancer, detection of single virus particles, and the detection of nitro-aromatic explosive materials such as 2,4,6 Tri-nitrotoluene (TNT) in sensitives superior to these of canines. Silicon nanowires could also be used in their twisted form, as electromechanical devices, to measure intermolecular forces with great precision.

Limitations of Sensing with Silicon Nanowire FET devices

Generally, the charges on dissolved molecules and macromolecules are screened by dissolved counterions, since in most cases molecules bound to the devices are separated from the sensor surface by approximately 2^{-12} nm (the size of the receptor proteins or

DNA linkers bound to the sensor surface). As a result of the screening, the electrostatic potential that arises from charges on the analyte molecule decays exponentially toward zero with distance. Thus, for optimal sensing, the Debye length must be carefully selected for nanowire FET measurements. One approach of overcoming this limitation employs fragmentation of the antibody-capturing units and control over surface receptor density, allowing more intimate binding to the nanowire of the target protein. This approach proved useful for dramatically enhancing the sensitivity of cardiac biomarkers (e.g., Troponin) detection directly from serum for the diagnosis of acute myocardial infarction.

NANORADIO

A nanoradio (also called carbon nanotube radio) is a nanotechnology acting as a radio transmitter and receiver by using carbon nanotubes. One of the first nanoradios was constructed in 2007 by researchers under Alex Zettl at the University of California, Berkeley where they successfully transmitted an audio signal. Due to the small size, nanoradios can have several possible applications such as radio function in the bloodstream.

The first observation of a nanoradio can be accredited to a Japanese physicist Sumio Iijima in 1991 who saw a "a luminous discharge of electricity" coming from a carbon nanotube on a graphite electrode. On October 31, 2007, a team of researchers under Alex Zettl at the University of California, Berkeley created one of the first nanoradios. Their experiment consisted of placing a multilayered nanotube placed on a silicon electrode and connecting it to a counter electrode through a wire and a DC battery. Both the electrode and nanotube were also put in a vacuum of about 10^{-7} Torr. They then placed the apparatus into a high-resolution transmission electron microscope to document the movement of the nanotube. They observed the nanoradio vibrating and transmitted a song called "Layla" by Eric Clapton. After some minor adjustments, the team was able to transmit and receive signals from a couple meters across the laboratory; however, the initial audio receptions from the radio were scratchy which Zettl believed was due to the lack of a better vacuum.

Properties

The small size, roughly 10 nanometers wide and hundreds of nanometers long, and composition of nanoradios provide several distinct properties. The small size of nanoradios enables electrons to pass through without much friction, making nanoradios efficient conductors. Nanoradios can also come in different sizes; they can be double-walled, tripled-walled and multi-walled. Aside from the different sizes, nanoradios can also take different shapes such as bent, straight or toroidal. Common among all nanoradios is how relatively strong they are. The resistance can be attributed to the strength of the bonds between carbon atoms.

Function

The fundamental parts of a radio are the antenna, tuner, demodulator and amplifier. Carbon nanotubes are special in that they can function as these parts without the need of extra circuitry.

Antenna

The nanoradio is small enough for electromagnetic signals to mechanically vibrate the nanoradio. The nanoradio essentially acts as an antenna by vibrating with the same frequency as the signal from incoming electromagnetic waves; this is in contrast with traditional radio antennas, which are generally stationary. The nanotube can vibrate in high frequencies, from "thousands to millions of times per second."

Tuner

The nanoradio can also function as a tuner by extending or reducing the length of the nanotube; doing so changes the resonance frequency at which it vibrates, enabling the radio to tune into specific frequencies. The length of the nanotube can be extended by pulling the tip with a positive electrode and can be shortened by removing atoms off the tip. Consequently, changing the length is permanent and can't be reversed; however, the method of varying the electric field can also affect the frequency that the nanoradio responds without being permanent.

Amplifier

As a benefit of the microscopic size and needle-like shape, the nanoradio functions naturally as an amplifier. The nanoradio exhibits field emission, in which a small voltage emits a flow of electrons; due to this, a small electromagnetic wave would produce a large flow of electrons, amplifying the signal.

Demodulator

Demodulation is essentially the separation of the information signal from the carrier wave. When the nanoradio vibrates in sync with the carrier wave, the nanoradio responds only to the information signal and ignores the carrier wave; and so, the nanoradio can act as a demodulator without the need of circuitry.

Medical Application

Currently, chemotherapy uses chemicals that harm not only cancerous cells, but also healthy ones since they are put into the blood stream. Nanoradios can be used to prevent damage to healthy cells by remotely communicating with the radio to release drugs and specifically target cancerous cells. Nanoradios can also be injected into individual cells to release certain chemicals, enabling repair of specific cells. Nanoradios can also be used to monitor insulin levels of diabetes patients and use that information to release a drug or chemical.

Complications

The implanting of nanoradios in the body are currently not feasible due to power dissipation. The nanoradio radiates about 4.5×10^{-27} W of electromagnetic power; however, much of this power would be lost when passing through the body. The amount of energy input can be increased, but would generate too much heat in the body, posing a safety risk. Other issues include the difficulty of constructing a nanoradio due to its nanoscale size, requiring quantum models and precision to manufacture.

NANOSENSOR

Nanosensors are nanoscale devices that measure physical quantities and convert those quantities to signals that can be detected and analyzed. There are several ways being proposed today to make nanosensors; these include top-down lithography, bottom-up assembly, and molecular self-assembly. There are different types of nanosensors in the market and in development for various applications. Though all sensors measure different things, sensors share the same basic workflow: a selective binding of an analyte, signal generation from the interaction of the nanosensor with the bio-element, and processing of the signal into useful metrics.

Nanomaterials-based sensors have several benefits in sensitivity and specificity over sensors made from traditional materials. Nanosensors can have increased specificity because they operate at a similar scale as natural biological processes, allowing functionalization with chemical and biological molecules, with recognition events that cause detectable physical changes. Enhancements in sensitivity stem from the high surface-to-volume ratio of nanomaterials, as well as novel physical properties of nanomaterials that can be used as the basis for detection, including nanophotonics. Nanosensors can also potentially be integrated with nanoelectronics to add native processing capability to the nanosensor.

In addition to their sensitivity and specificity, nanosensors offer significant advantages in cost and response times, which makes nanosensors suitable for high-throughput applications. Nanosensors provide real-time monitoring compared to traditional detection methods such as chromatography and spectroscopy. These traditional methods may take days to weeks to obtain results and often require investment in capital costs as well as time for sample preparation.

One-dimensional nanomaterials such as nanowires and nanotubes are well suited for use in nanosensors, as compared to bulk or thin-film planar devices. They can function both as transducers and wires to transmit the signal. Their high surface area can cause large signal changes upon binding of an analyte. Their small size can enable extensive multiplexing of individually addressable sensor units in a small device. Their operation is also "label free" in the sense of not requiring fluorescent or radioactive labels on the analytes.

There are several challenges for nanosensors, including avoiding fouling and drift, developing reproducible calibration methods, applying preconcentration and separation methods to attain a proper analyte concentration that avoids saturation, and integrating the nanosensor with other elements of a sensor package in a reliable manufacturable manner. Because nanosensors are a relatively new technology, there are many unanswered questions regarding nanotoxicology which currently limits their application in biological systems. Some nanosensors may impact cell metabolism and homeostasis, changing cellular molecular profiles and making it difficult to separate sensor-induced artifacts from fundamental biological phenomena.

Potential applications for nanosensors include medicine, detection of contaminants and pathogens, and monitoring manufacturing processes and transportation systems. By measuring changes in physical properties (volume, concentration, displacement and velocity, gravitational, electrical, and magnetic forces, pressure, or temperature) nanosensors may be able to distinguish between and recognize certain cells at the molecular level in order to deliver medicine or monitor development to specific places in the body. The type of signal transduction defines the major classification system for nanosensors. Some of the main types of nanosensor readouts include optical, mechanical, vibrational, or electromagnetic.

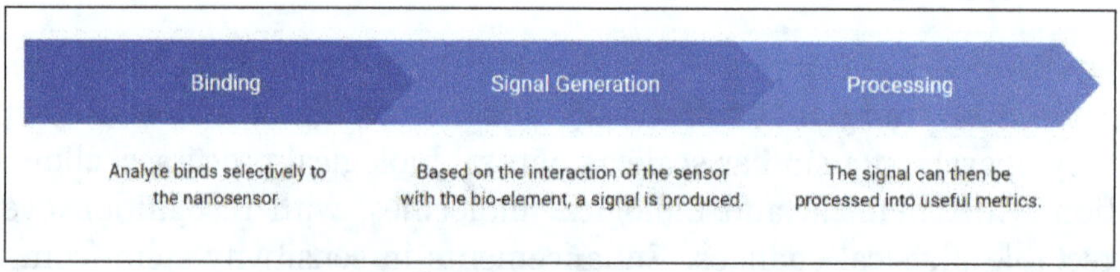

Overview of a general nanosensor workflow.

Mechanisms of Operation

There are many mechanisms by which a recognition event can be transduced into a measurable signal. Electrochemical nanosensors are based on detecting a resistance change in the nanomaterial upon binding of an analyte, due to changes in scattering or to the depletion or accumulation of charge carriers. One possibility is to use nanowires such as carbon nanotubes, conductive polymers, or metal oxide nanowires as gates in field-effect transistors, although as of 2009 they had not yet been demonstrated in real-world conditions. Chemical nanosensors contain a chemical recognition system (receptor) and a physiochemical transducer, in which the receptor interacts with analyte to produce electrical signals. In one case, upon interaction of the analyte with the receptor, the nanoporous transducer had a change in impedance which was determined as the sensor signal. Other examples include electromagnetic or plasmonic nanosensors, spectroscopic nanosensors such as surface-enhanced Raman spectroscopy, magnetoelectronic or spintronic nanosensors, and mechanical nanosensors.

Photonic devices can also be used as nanosensors to quantify concentrations of clinically relevant samples. A principle of operation of these sensors is based on the chemical modulation of a hydrogel film volume that incorporates a Bragg grating. As the hydrogel swells or shrinks upon chemical stimulation, the Bragg grating changes color and diffracts light at different wavelengths. The diffracted light can be correlated with the concentration of a target analyte.

Another type of nanosensor is one that works through a colorimetric basis. Here, the presence of the analyte causes a chemical reaction or morphological alteration for a visible color change to occur. One such application, is that gold nanoparticles can be used for the detection of heavy metals. Many harmful gases can also be detected by a colorimetric change, such as through the commercially available Dräger Tube. These provide an alternative to bulky, lab-scale systems, as these can be miniaturized to be used for point-of-sample devices. For example, many chemicals are regulated by the Environmental Protection Agency and require extensive testing to ensure contaminant levels are within the appropriate limits. Colorimetric nanosensors provide a method for on-site determination of many contaminants.

Production Methods

There are currently several hypothesized ways to produce nanosensors. Top-down lithography is the manner in which most integrated circuits are now made. It involves starting out with a larger block of some material and carving out the desired form. These carved out devices, notably put to use in specific microelectromechanical systems used as microsensors, generally only reach the micro size, but the most recent of these have begun to incorporate nanosized components.

Another way to produce nanosensors is through the bottom-up method, which involves assembling the sensors out of even more minuscule components, most likely individual atoms or molecules. This would involve moving atoms of a particular substance one by one into particular positions which, though it has been achieved in laboratory tests using tools such as atomic force microscopes, is still a significant difficulty, especially to do en masse, both for logistic reasons as well as economic ones. Most likely, this process would be used mainly for building starter molecules for self-assembling sensors.

The third way, which promises far faster results, involves self-assembly, or "growing" particular nanostructures to be used as sensors. This most often entails an already complete set of components that would automatically assemble themselves into a finished product. Accurately being able to reproduce this effect for a desired sensor in a laboratory would imply that scientists could manufacture nanosensors much more quickly and potentially far more cheaply by letting numerous molecules assemble themselves with little or no outside influence, rather than having to manually assemble each sensor.

Applications

One of the first working examples of a synthetic nanosensor was built by researchers at the Georgia Institute of Technology in 1999. It involved attaching a single particle onto the end of a carbon nanotube and measuring the vibrational frequency of the nanotube both with and without the particle. The discrepancy between the two frequencies allowed the researchers to measure the mass of the attached particle.

Since then, increasing amounts of research have gone into nanosensors, whereby modern nanosensors have been developed for many applications. Currently, the applications of nanosensors in the market include: healthcare, defense and military, and others such as food, environment, and agriculture.

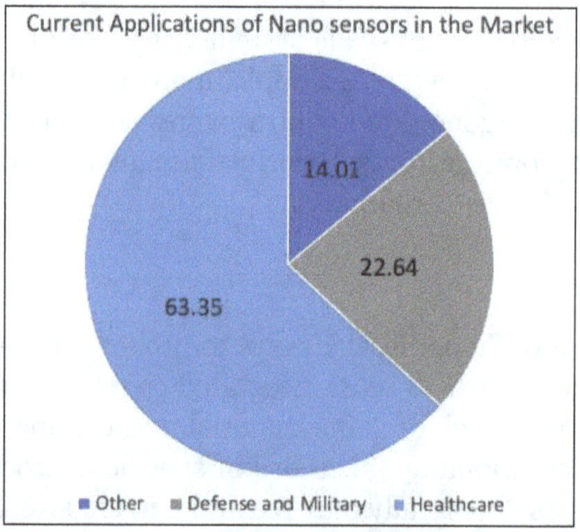

Brief breakdown of current industry applications of nanosensors.

Defense and Military

Nanoscience as a whole has much potential for applications in the defense and military sector. Applications include chemical detection, decontamination, and forensics. However, the primary efforts surrounding nanosensors largely remain in research and development.

Some nanosensors in development for defense applications include nanosensors for the detection of explosives or toxic gases. Such nanosensors work on the principle that gas molecules can be distinguished based on their mass using, for example, piezoelectric sensors. If a gas molecule is adsorbed at the surface of the detector, the resonance frequency of the crystal changes and this can be measured as a change in electrical properties. In addition, field effect transistors, used as potentiometers, can detect toxic gases if their gate is made sensitive to them.

In a similar application, nanosensors can be utilized in military and law enforcement clothing and gear. The Navy Research Laboratory's Institute for Nanoscience has studied quantum dots for application in nanophotonics and identifying biological materials. Nanoparticles layered with polymers and other receptor molecules will change color when contacted by analytes such as toxic gases. This alerts the user that they are in danger. Other projects involve embedding clothing with biometric sensors to relay information regarding the user's health and vitals, which would be useful for monitoring soldiers in combat.

Surprisingly, some of the most challenging aspects in creating nanosensors for defense and military use are political in nature, rather than technical. Many different government agencies must work together to allocate budgets and share information and progress in testing; this can be difficult with such large and complex institutions. In addition, visas and immigration status can become an issue for foreign researchers - as the subject matter is very sensitive, government clearance can sometimes be required. Finally, there are currently not well defined or clear regulations on nanosensor testing or applications in the sensor industry, which contributes to the difficulty of implementation.

Food and the Environment

Nanosensors can improve various sub-areas within food and environment sectors including food processing, agriculture, air and water quality monitoring, and packaging and transport. These nanosensors enable rapid analysis of samples and detection of contaminants in food which is especially important for downstream processing and perishable items.

Chemical sensors are useful for analyzing odors from food samples and detecting atmospheric gases. The "electronic nose" was developed in 1988 to determine the quality and freshness of food samples using traditional sensors, but more recently the sensing film has been improved with nanomaterials. A sample is placed in a chamber where volatile compounds become concentrated in the gas phase, whereby the gas is then pumped through the chamber to carry the aroma to the sensor that measures its unique fingerprint. The high surface area to volume ratio of the nanomaterials allows for greater interaction with analytes and the nanosensor's fast response time enables the separation of interfering responses. Chemical sensors, too, have been built using nanotubes to detect various properties of gaseous molecules. Carbon nanotubes have been used to sense ionization of gaseous molecules while nanotubes made out of titanium have been employed to detect atmospheric concentrations of hydrogen at the molecular level. Many of these involve a system by which nanosensors are built to have a specific pocket for another molecule. When that particular molecule, and only that specific molecule, fits into the nanosensor, and light is shone upon the nanosensor, it will reflect different wavelengths of light and, thus, be a different color. In a similar fashion, Flood have shown that supramolecular host–guest chemistry offers quantitative sensing using Raman scattered light as well as SERS.

Other types of nanosensors, including quantum dots and gold nanoparticles, are currently being developed to detect pollutants and toxins in the environment. Quantum dot surfaces can be modified with antibodies that bind specifically to environmental pollutants. Gold nanoparticle-based optical sensors can be used to detect heavy metals very precisely; for example, mercury levels as low as 0.49 nM. This sensing modality takes advantage of fluorescence resonance energy transfer (FRET), in which the presence of metals inhibits the interaction between quantum dots and gold nanoparticles, and quenches the FRET response.

The main challenge associated with using nanosensors in food and the environment is determining their associated toxicity and overall effect on the environment. Currently, there is insufficient knowledge on how the implementation of nanosensors will affect the soil, plants, and humans in the long-term. This is difficult to fully address because nanoparticle toxicity depends heavily on the type, size, and dosage of the particle as well as environmental variables including pH, temperature, and humidity. To mitigate potential risk, research is being done to manufacture safe, nontoxic nanomaterials, as part of an overall effort towards green nanotechnology.

Healthcare

One example of nanosensors involves using the fluorescence properties of cadmium selenide quantum dots as sensors to uncover tumors within the body. A downside to the cadmium selenide dots, however, is that they are highly toxic to the body. As a result, researchers are working on developing alternate dots made out of a different, less toxic material while still retaining some of the fluorescence properties. In particular, they have been investigating the particular benefits of zinc sulfide quantum dots which, though they are not quite as fluorescent as cadmium selenide, can be augmented with other metals including manganese and various lanthanide elements. In addition, these newer quantum dots become more fluorescent when they bond to their target cells.

Another application of nanosensors involves using silicon nanowires in IV lines to monitor organ health. The nanowires are sensitive to detect trace biomarkers that diffuse into the IV line through blood which can monitor kidney or organ failure. These nanowires would allow for continuous biomarker measurement, which provides some benefits in terms of temporal sensitivity over traditional biomarker quantification assays such as ELISA.

Nanosensors can also be used to detect contamination in organ implants. The nanosensor is embedded into the implant and detects contamination in the cells surrounding the implant through an electric signal sent to a clinician or healthcare provider. The nanosensor can detect whether the cells are healthy, inflammatory, of contaminated with bacteria.

Currently, there are stringent regulations in place for the development of standards for nanosensors to be used in the medical industry, due to insufficient knowledge of the adverse effects of nanosensors as well as potential cytotoxic effects of nanosensors.

Additionally, there can be a high cost of raw materials such as silicon, nanowires, and carbon nanotubes, which prevent commercialization and manufacturing of nanosensors requiring scale-up for implementation. To mitigate the drawback of cost, researchers are looking into manufacturing nanosensors made of more cost-effective materials. There is also a high degree of precision needed to reproducibly manufacture nanosensors, due to their small size and sensitivity to different synthesis techniques, which creates additional technical challenges to be overcome.

NANOPHOTONICS

Nanophotonics or nano-optics is the study of the behavior of light on the nanometer scale, and of the interaction of nanometer-scale objects with light. It is a branch of optics, optical engineering, electrical engineering, and nanotechnology. It often (but not exclusively) involves metallic components, which can transport and focus light via surface plasmon polaritons.

The term "nano-optics", just like the term "optics", usually refers to situations involving ultraviolet, visible, and near-infrared light (free-space wavelengths from 300 to 1200 nanometers).

Normal optical components, like lenses and microscopes, generally cannot normally focus light to nanometer (deep subwavelength) scales, because of the diffraction limit (Rayleigh criterion). Nevertheless, it is possible to squeeze light into a nanometer scale using other techniques like, for example, surface plasmons, localized surface plasmons around nanoscale metal objects, and the nanoscale apertures and nanoscale sharp tips used in near-field scanning optical microscopy (SNOM or NSOM) and photoassisted scanning tunnelling microscopy.

Motivations

Nanophotonics researchers pursue a very wide variety of goals, in fields ranging from biochemistry to electrical engineering. A few of these goals are summarized below.

Optoelectronics and Microelectronics

If light can be squeezed into a small volume, it can be absorbed and detected by a small detector. Small photodetectors tend to have a variety of desirable properties including low noise, high speed, and low voltage and power.

Small lasers have various desirable properties for optical communication including low threshold current (which helps power efficiency) and fast modulation (which means more data transmission). Very small lasers require subwavelength optical cavities. An example is spasers, the surface plasmon version of lasers.

Integrated circuits are made using photolithography, i.e. exposure to light. In order to make very small transistors, the light needs to be focused into extremely sharp images. Using various techniques such as immersion lithography and phase-shifting photomasks, it has indeed been possible to make images much finer than the wavelength—for example, drawing 30 nm lines using 193 nm light. Plasmonic techniques have also been proposed for this application.

Heat-assisted magnetic recording is a nanophotonic approach to increasing the amount of data that a magnetic disk drive can store. It requires a laser to heat a tiny, subwavelength area of the magnetic material before writing data. The magnetic write-head would have metal optical components to concentrate light at the right location.

Miniaturization in optoelectronics, for example the miniaturization of transistors in integrated circuits, has improved their speed and cost. However, optoelectronic circuits can only be miniaturized if the optical components are shrunk along with the electronic components. This is relevant for on-chip optical communication (i.e. passing information from one part of a microchip to another by sending light through optical waveguides, instead of changing the voltage on a wire).

Solar Cells

Solar cells often work best when the light is absorbed very close to the surface, both because electrons near the surface have a better chance of being collected, and because the device can be made thinner, which reduces cost. Researchers have investigated a variety of nanophotonic techniques to intensify light in the optimal locations within a solar cell.

Spectroscopy

Using nanophotonics to create high peak intensities: If a given amount of light energy is squeezed into a smaller and smaller volume ("hot-spot"), the intensity in the hot-spot gets larger and larger. This is especially helpful in nonlinear optics; an example is surface-enhanced Raman scattering. It also allows sensitive spectroscopy measurements of even single molecules located in the hot-spot, unlike traditional spectroscopy methods which take an average over millions or billions of molecules.

Microscopy

One goal of nanophotonics is to construct a so-called "superlens", which would use metamaterials or other techniques to create images that are more accurate than the diffraction limit (deep subwavelength).

Near-field scanning optical microscope (NSOM or SNOM) is a quite different nanophotonic technique that accomplishes the same goal of taking images with resolution far smaller than the wavelength. It involves raster-scanning a very sharp tip or very small aperture over the surface to be imaged.

Near-field microscopy refers more generally to any technique using the near-field to achieve nanoscale, subwavelength resolution. For example, dual-polarization interferometry has picometer resolution in the vertical plane above the waveguide surface.

Principles

Plasmons and Metal Optics

Metals are an effective way to confine light to far below the wavelength. This was originally used in radio and microwave engineering, where metal antennas and waveguides may be hundreds of times smaller than the free-space wavelength. For a similar reason, visible light can be confined to the nano-scale via nano-sized metal structures, such as nano-sized structures, tips, gaps, etc. Many nano-optics designs look like common microwave or radiowave circuits, but shrunk down by a factor of 100,000 or more. After all, radiowaves, microwaves, and visible light are all electromagnetic radiation; they differ only in frequency. So other things equal, a microwave circuit shrunk down by a factor of 100,000 will behave the same way but at 100,000 times higher frequency. This effect is somewhat analogous to a lightning rod, where the field concentrates at the tip. It is fundamentally based on the fact that the permittivity of the metal is very large and negative. At very high frequencies (near and above the plasma frequency, usually ultraviolet), the permittivity of a metal is not so large, and the metal stops being useful for concentrating fields.

For example, researchers have made nano-optical dipoles and Yagi–Uda antennas following essentially the same design as used for radio antennas.

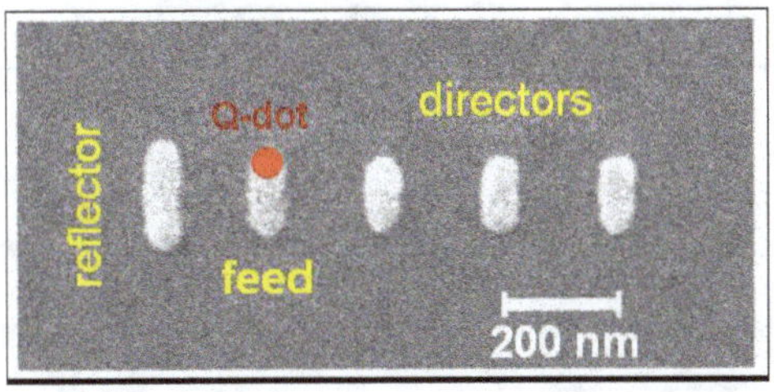

Scanning electron microscopy (SEM) image of a five-element Yagi-Uda antenna consisting of a feed element, one reflector, and three directors, fabricated by e-beam lithography.

Metallic parallel-plate waveguides (striplines), lumped-constant circuit elements such as inductance and capacitance (at visible light frequencies, the values of the latter being of the order of femtohenries and attofarads, respectively), and impedance-matching of dipole antennas to transmission lines, all familiar techniques at microwave frequencies, are some current areas of nanophotonics development. That said, there are a number of very important differences between nano-optics and scaled-down microwave circuits. For example, at optical frequency, metals behave much less like ideal conductors, and also exhibit interesting plasmon-related effects like kinetic inductance and surface plasmon resonance. Likewise, optical fields interact with semiconductors in a fundamentally different way than microwaves do.

Near-field Optics

If you take the Fourier transform of an object, it consists of different spatial frequencies. The higher frequencies correspond to the very fine features and sharp edges.

When light is emitted by such an object, the light with very high spatial frequency forms an evanescent wave, which only exists in the near field (very close to the object, within a wavelength or two) and disappears in the far field. This is the origin of the diffraction limit, which says that when a lens images an object, the subwavelength information is blurred out.

Nano-photonics is primarily concerned with the near-field evanescent waves. For example, a superlens would prevent the decay of the evanescent wave, allowing higher-resolution imaging.

Metamaterials

Metamaterials are artificial materials engineered to have properties that may not be found in nature. They are created by fabricating an array of structures much smaller than a wavelength. The small (nano) size of the structures is important: That way, light interacts with them as if they made up a uniform, continuous medium, rather than scattering off the individual structures.

NANOIONICS

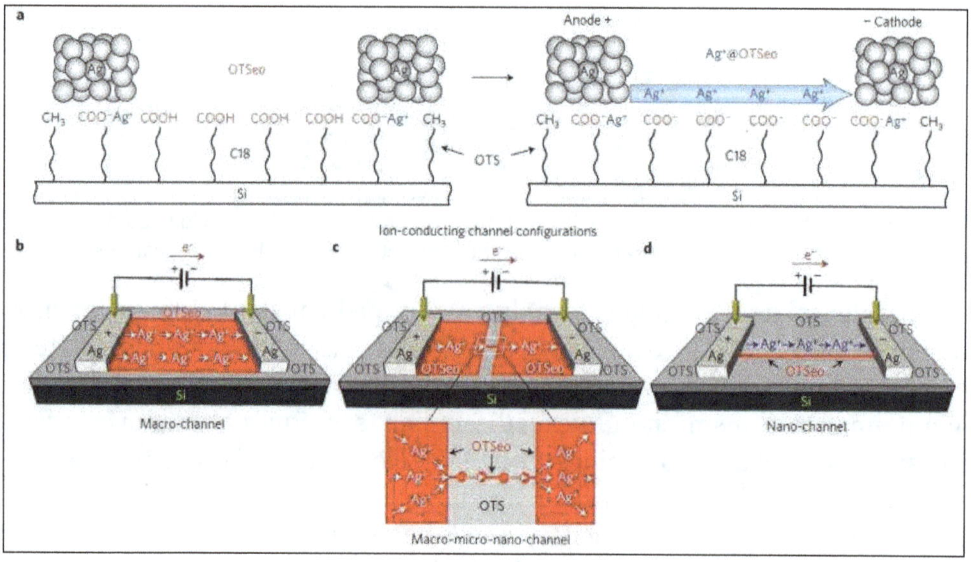

Nanoionics is a sub-field of nanotechnology that is concerned with nanoscale phenomena involving migration of ions in solid maters. Thus far, however, the confinement of ionic currents to predefined paths in a manner resembling the motion of electrons in wires of electronic conductors has not been explored.

Constructive lithography involves an electrochemical oxidative reaction between a conductive atomic force microscope (AFM) tip and the molecules attached to a silicon wafer, typically as monolayers. This allows for a level of precision in which one can select which molecules will undergo the reaction and which ones will not. In this study, oganosilane monolayers comprised of $Si\text{-}Cl_3$ anchors, an aliphatic carbon backbone, and a methyl ($-CH_3$) terminal group are selectively oxidized using constructive lithography. The methyl terminal group is oxidized to a carboxylic acid terminal group (-COOH), without changing the backbone and anchor.

Constructive lithography allows for the creation of sharp boundary points. These boundaries are between surface regions populated by methyl-terminated molecules and those populated by carboxylic acid-terminated molecules. Electrodes of a given metal can be placed on the carboxylic acid-terminated surface regions at such boundary sites, thus setting up a carboxylic pathway for ions to travel.

For the initial model system in this research, Berson and Burshtain, placed silver (Ag) electrodes at boundary sites, and then ran d.c. voltage through the electrodes, thus producing mobile silver ions. Their goal was to see if the Ag+ ions would coordinate to the deprotonated carboxylic acids, essentially traversing the length of the specified pathway from anode to cathode without the use of an added electrolyte.

Fourier Transform Infrared spectroscopy (FTIR) verified that the targeted terminal $-CH_3$ groups converted to –COOH without disturbing the other segments of the monolayer. Furthermore, FTIR readings were taken over a period of time while a d.c. voltage of 1 mV to 100 mV was applied. This confirmed that –COOH lost its proton forming the silver carboxylate salt. XPS confirmed that the silver ions followed the carboxylate pathway located between the two electrodes and did not interact with the methyl-terminated molecules. Resistance studies elucidated the transport of ions across channels of varying lengths and widths.

The same procedure was followed with titanium electrodes instead of silver electrodes. Berson and Burshtain looked at a system with titanium electrodes serving as the anode and the cathode, and a combination system with silver and titanium electrodes. FTIR confirmed the presence of Ti^{4+} on the surface of the system with titanium electrodes. The mixed electrode system showed evidence that one cation can replace the other one on the carboxylate surface while competing for available $-COO^-$ sites.

Arrhenius plots of the silver and titanium system indicate that a similar conduction mechanism occurs in the mixed metal system compared to the single metal system. There is a difference between titanium's activation energy compared to silver, which the authors believe is likely due to Ti^{4+} coordinating to four $-COO^-$ molecules compared to Ag^+ coordinating to one. Furthermore, Ti^{4+} bonding takes on a slightly more covalent character than Ag^+.

While the aforementioned studies were done at the macroscale, the next step was to see if the model system would work with nano-channel configurations. The authors constructed a system with two macro-channels that were separated by a nano-channel with silver electrodes residing in the macro-channels and another system in which the silver electrodes were on either side of a nano-channel.

Resistance calculations and AFM images indicated that the system with two macro-channels separated by a nano-channel behaved similarly to the macro-channels. However, when the electrodes were placed on the boundaries of the nano-channel, a thin film of silver accumulated along the channel instead of at the cathode, likely due to higher flux density of ions in the system and therefore more opportunity for nucleation and subsequent silver filament growth along the nano-channel pathway.

This research demonstrates the ability to tailor ion-conducting conductive channels using constructive lithography on alkyl silane monolayers. This technique is versatile in that the conductive channels may accommodate different mobile ions produced by using different metals can be used for the electrodes.

References

- Nanoelectronics: circuitstoday.com, Retrieved 08 January, 2020
- D. Mihailovic (2009). "Inorganic molecular wires: Physical and functional properties of transition metal chalco-halide polymers". Progress in Materials Science. 54 (3): 309–350. doi:10.1016/j.pmatsci.2008.09.001
- Nanoionics-versatile-ion-conducting-channels-monolayers: phys.org, Retrieved 10 March, 2020
- Waser, Rainer; Lüssem, B. & Bjørnholm, T. (2008). "Chapter 8: Concepts in Single-Molecule Electronics". Nanotechnology. Volume 4: Information Technology II. Weinheim: Wiley-VCH Verlag GmbH & Co. KGaA. pp. 175–212. ISBN 978-3-527-31737-0
- Spencer, Joseph; Nesbitt, John; Trewhitt, Harrison; Kashtiban, Reza; Bell, Gavin; Ivanov, Victor; Faulques, Eric; Smith, David (2014). "Raman Spectroscopy of Optical Transitions and Vibrational Energies of ~1 nm HgTe Extreme Nanowires within Single Walled Carbon Nanotubes" (PDF). ACS Nano. 8 (9): 9044–52. doi:10.1021/nn5023632. PMID 25163005
- Acuna, Guillermo; Grohmann, Dina; Tinnefeld, Philip (2014). "Enhancing single-molecule fluorescence with nanophotonics". FEBS Letters. 588 (19): 3547–3552. doi:10.1016/j.febslet.2014.06.016. PMID 24928436

5

Applications

Nanotechnology has varied applications in the fields of food science, medicine, robotics, architecture, defense, space technology, industry, agriculture, etc. This chapter closely examines these applications of nanotechnology to provide an extensive understanding of the subject.

APPLICATION OF NANOTECHNOLOGY IN FOOD SCIENCE

Recent innovations in nanotechnology have transformed a number of scientific and industrial areas including the food industry. Applications of nanotechnology have emerged with increasing need of nanoparticle uses in various fields of food science and food microbiology, including food processing, food packaging, functional food development, food safety, detection of foodborne pathogens, and shelf-life extension of food and/or food products.

Over the past few decades, nanotechnology has increasingly been considered as to be attractive technology that has revolutionized the food sector. It is a technology on the nanometer scale and deals with the atoms, molecules, or the macromolecules with the size of approximately 1–100 nm to create and use materials that have novel properties. The created nanomaterials possess one or more external dimensions, or an internal structure, on the scale from 1 to 100 nm that allowed the observation and manipulation of matter at the nanoscale. It is observed that these materials have unique properties unlike their macroscale counterparts due to the high surface to volume ratio and other novel physiochemical properties like color, solubility, strength, diffusivity, toxicity, magnetic, optical, thermodynamic, etc.. Nanotechnology has brought new industrial revolution and both developed and developing countries are interested in investing more in this technology. Therefore, nanotechnology offers a wide range of opportunities for the development and application of structures, materials, or system with new properties in various areas like agriculture, food, and medicine, etc.

The rising consumer concerns about food quality and health benefits are impelling the researchers to find the way that can enhance food quality while disturbing least the nutritional value of the product. The demand of nanoparticle-based materials has been increased in the food industry as many of them contain essential elements and also found to be non-toxic. They have been also found to be stable at high temperature and pressures. Nanotechnology offers complete food solutions from food manufacturing, processing to packaging. Nanomaterials bring about a great difference not only in the food quality and safety but also in health benefits that food delivers. Many organizations, researchers, and industries are coming up with novel techniques, methods, and products that have a direct application of nanotechnology in food science.

The applications of nanotechnology in food sector can be summarized in two main groups that are food nanostructured ingredients and food nanosensing. Food nanostructured ingredients encompass a wide area from food processing to food packaging. In food processing, theses nanostructures can be used as food additives, carriers for smart delivery of nutrients, anti-caking agents, antimicrobial agents, fillers for improving mechanical strength and durability of the packaging material, etc. whereas food nanosensing can be applied to achieve better food quality and safety evaluation.

Nanotechnology in Food Processing

The nanostructured food ingredients are being developed with the claims that they offer improved taste, texture, and consistency. Nanotechnology increasing the shelf-life of different kinds of food materials and also help brought down the extent of wastage of food due to microbial infestation. Nowadays nanocarriers are being utilized as delivery systems to carry food additives in food products without disturbing their basic morphology. Particle size may directly affect the delivery of any bioactive compound to various sites within the body as it was noticed that in some cell lines, only submicron nanoparticles can be absorbed efficiently but not the larger size micro-particles. An ideal delivery system is supposed to have following properties: (i) able to deliver the active compound precisely at the target place (ii) ensure availability at a target time and specific rate, and (iii) efficient to maintain active compounds at suitable levels for long periods of time (in storage condition). Nanotechnology being applied in the formation of encapsulation, emulsions, biopolymer matrices, simple solutions, and association colloids offers efficient delivery systems with all the above-mentioned qualities. Nano polymers are trying to replace conventional materials in food packaging. Nanosensors can be used to prove the presence of contaminants, mycotoxins, and microorganisms in food.

Nanoparticles have better properties for encapsulation and release efficiency than traditional encapsulation systems. Nanoencapsulations mask odors or tastes, control interactions of active ingredients with the food matrix, control the release of the active agents, ensure availability at a target time and specific rate, and protect them from moisture, heat, chemical, or biological degradation during processing, storage, and utilization, and also exhibit compatibility with other compounds in the system. Moreover, these delivery systems possess the ability to penetrate deeply into tissues due to their smaller size and thus allow efficient delivery of active compounds to target sites in the body. Various synthetic and natural polymer-based encapsulating delivery systems have been elaborated for the improved bioavailability and preservation of the active food components. Further, the importance of nanotechnology in food processing can be evaluated by considering its role in the improvement of food products in terms of (i) food texture, (ii) food appearance, (iii) food taste, (iv) nutritional value of the food, and (v) food shelf-life. It is a matter of fact that surprisingly nanotechnology not only touches all the above-mentioned aspects but has also brought about significant alterations in food products providing them novel qualities.

Texture, Taste and Appearance of Food

Nanotechnology provides a range of options to improve the food quality and also helps in enhancing food taste. Nanoencapsulation techniques have been used broadly to improve the flavor release and retention and to deliver culinary balance. Zhang used the nanoencapsulation for the highly reactive and unstable plant pigment anthocyanins which have various biological activities. Through, encapsulating cyanidin-3-O-glucoside (C3G) molecules within the inner cavity of apo recombinant soybean seed H-2 subunit ferritin (rH-2) improved the thermal stability and photostability. This design and fabrication of multifunctional nanocarriers for bioactive molecule protection and delivery. Rutin is a common dietary flavonoid with great important pharmacological activities but due to poor solubility, its application in the food industry is limited. The ferritin nanocages encapsulation enhanced the solubility, thermal and UV radiation stability of ferritin trapped rutin as compared to free rutin. The use of nanoemulsions to deliver lipid-soluble bioactive compounds is much popular since they can be produced using natural food ingredients using easy production methods, and may be designed to enhance water-dispersion and bioavailability.

As compared to larger particles which generally release encapsulated compounds more slowly and over longer time periods, nanoparticles provide promising means of improving the bioavailability of nutraceutical compounds due to their subcellular size leading to a higher drug bioavailability. Many metallic oxides such as titanium dioxide and silicon dioxide (SiO_2) have conventionally been used as color or flow agents in food items. SiO_2 nanomaterials are also one of the most used food nanomaterials as carriers of fragrances or flavors in food products.

Nutritional Value

A majority of bioactive compounds such as lipids, proteins, carbohydrates, and vitamins are sensitive to high acidic environment and enzyme activity of the stomach and duodenum. Encapsulation of these bioactive compounds not only enables them to resist such adverse conditions but also allows them to assimilate readily in food products, which is quite hard to achieve in non-capsulated form due to low water-solubility of these bioactive compounds. Nanoparticles-based tiny edible capsules with the aim to improve delivery of medicines, vitamins or fragile micronutrients in the daily foods are being created to provide significant health benefits. The nanocomposite, nano-emulsification, and nanostructuration are the different techniques which have been applied to encapsulate the substances in miniature forms to more effectively deliver nutrients like protein and antioxidants for precisely targeted nutritional and health benefits. Polymeric nanoparticles are found to be suitable for the encapsulation of bioactive compounds (e.g., flavonoids and vitamins) to protect and transport bioactive compounds to target functions.

Preservation or Shelf-life

In functional foods where bioactive component often gets degraded and eventually led to inactivation due to the hostile environment, nanoencapsulation of these bioactive components extends the shelf-life of food products by slowing down the degradation processes or prevents degradation until the product is delivered at the target site. Moreover, the edible nano-coatings on various food materials could provide a barrier to moisture and gas exchange and deliver colors, flavors, antioxidants, enzymes, and anti-browning agents and could also increase the shelf-life of manufactured foods, even after the packaging is opened. Encapsulating functional components within the droplets often enables a slowdown of chemical degradation processes by engineering the properties of the interfacial layer surrounding them. For example, curcumin the most active and least stable bioactive component of turmeric (Curcuma longa) showed reduced antioxidant activity and found to be stable to pasteurization and at different ionic strength upon encapsulation.

Nanotechnology in Food Packaging

A desirable packaging material must have gas and moisture permeability combined with strength and biodegradability. Nano-based “smart” and “active” food packagings confer several advantages over conventional packaging methods from providing better packaging material with improved mechanical strength, barrier properties, antimicrobial films to nanosensing for pathogen detection and alerting consumers to the safety status of food.

Application of nanocomposites as an active material for packaging and material coating can also be used to improve food packaging. Many researchers were interested in studying the antimicrobial properties of organic compounds like essential oils, organic acids, and bacteriocins and their use in polymeric matrices as antimicrobial packaging. However, these compounds do not fit into the many food processing steps which require high temperatures and pressures as they are highly sensitive to these physical conditions. Using inorganic nanoparticles, a strong antibacterial activity can be achieved in low concentrations and more stability in extreme conditions. Therefore, in recent years, it has been a great interest of using these nanoparticles in antimicrobial food packaging.

An antimicrobial packaging is actually a form of active packaging which contacts with the food product or the headspace inside to inhibit or retard the microbial growth that may be present on food surfaces. Many nanoparticles such as silver, copper, chitosan, and metal oxide nanoparticles like titanium oxide or zinc oxide have been reported to have antibacterial property.

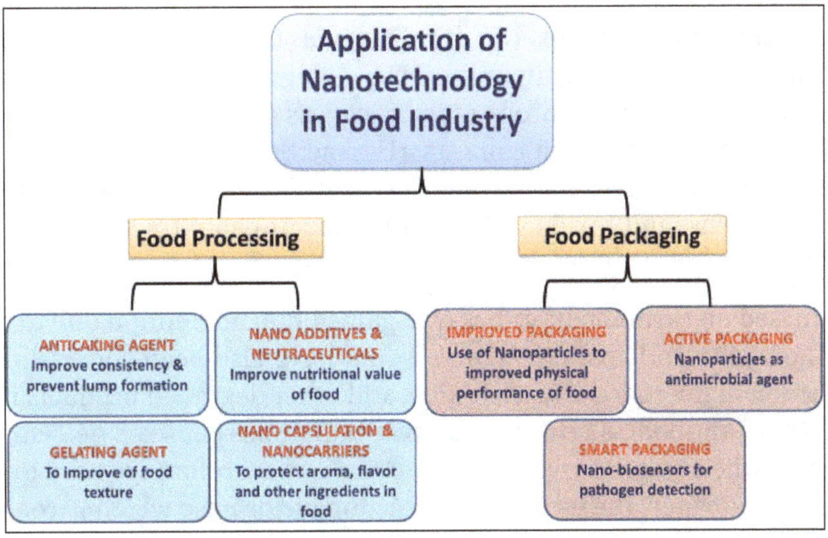

Schematic diagram showing role of nanotechnology in different aspects of food sectors.

The application of nanoparticles is not limited to antimicrobial food packaging but nano-composite and nanolaminates have been actively used in food packaging to provide a barrier from extreme thermal and mechanical shock extending food shelf-life. In this way, the incorporation of nanoparticles into packaging materials offers quality food with longer shelf-life. The purpose of creating polymer composites is to have more mechanical and thermostable packing materials. Many inorganic or organic fillers are being used in order to achieve improved polymer composites. The incorporation of nanoparticles in polymers has allowed developing more resist packaging material with cost effectiveness. Use of inert nanoscale fillers such as clay and silicate nanoplatelets, silica (SiO_2) nanoparticles, chitin or chitosan into the polymer matrix renders it lighter, stronger, fire resistance, and better thermal properties. Antimicrobial nanocomposite films which are prepared by impregnating the fillers (having at least one dimension in the nanometric range or nanoparticles) into the polymers offer two-way benefit because of their structural integrity and barrier properties.

Nanosensors for Pathogen Detection

Nanomaterials for use in the construction of biosensors offers the high level of sensitivity and other novel attributes. In food microbiology, nanosensors or nanobiosensors are used for the detection of pathogens in processing plants or in food material, quantification of available food constituents, alerting consumers and distributors on the safety status of food. The nanosensor works as an indicator that responds to changes in environmental conditions such as humidity or temperature in storage rooms, microbial contamination, or products degradation. Various nanostructures like thin films, nanorods, nanoparticles and nanofibers have been examined to their possible applications in biosensors. Thin film-based optical immunosensors for detection of microbial substances or cells have led to the rapid and highly sensitive detection systems. In these immunosensors, specific antibodies, antigens, or protein molecules are immobilized on thin nano-films or sensor chips which emit signals on detection of target molecules. A dimethylsiloxane microfluidic immunosensor integrated with specific antibody immobilized on an alumina nanoporous membrane was developed for rapid detection of foodborne pathogens Escherichia coli O157:H7 and Staphylococcus aureus with electrochemical impedance spectrum. Nanotechnology can also assist in the detection of pesticides, pathogens, and toxins serving in the food quality tracking–tracing–monitoring chain.

Biosensors based on carbon nanotubes also gained much attention due to their rapid detection, simplicity and cost effectiveness and have also been successfully applied for the detection of microorganisms, toxins, and other degraded products in food and beverages. Toxin antibodies attached to these nanotubes causes a detectable change in conductivity when bound to waterborne toxins and therefore are used to detect waterborne toxins. Further, the use of electronic tongue or nose which is consists of the array of nanosensors monitor the food condition by giving signals on aroma or gases released by food items. The quartz crystal microbalance (QCM)-based electric nose can detect the interaction between various odorants and chemicals that have been coated on the crystal surface of the QCM. Many studies on small molecule detection have used quartz crystal surfaces that have been modified with different functional groups or biological molecules, such as amines, enzymes, lipids, and various polymers.

Safety Issues

Besides a lot of advantages of nanotechnology to the food industry, safety issues associated with the nanomaterial cannot be neglected. Many researchers discussed safety concerns associated with nanomaterial giving emphasis on the possibility of nanoparticles migrate from the packaging material into the food and their impact on consumer's health. Although a material is being considered as GRAS (generally regarded as safe) substance, additional studies must be acquired to examine the risk of its nano counterparts because the physiochemical properties in nanostates are completely different from that are in macrostate. Moreover, the small size of these nanomaterials may increase the risk for bioaccumulation within body organs and tissues. For example, silica nanoparticles which are used as anti-caking agents can be cytotoxic in human lung cells when subjected to exposure. There are a lot of factors that affect dissolution including surface morphology of the particles, concentration, surface energy, aggregation, and adsorption. A model to study the migration of particles from food packaging has been developed by Cushen. They studied the migration of silver and copper from nanocomposites and observed that the percentage of nanofiller in the nanocomposites was one of the most crucial parameters driving migration, more so than particle size, temperature, or contact time. Since every nanomaterial has its individual property, therefore, toxicity will likely be established on a case-by-case basis. Further, regulatory authorities must develop some standards for commercial products to ensure product quality, health and safety, and environmental regulations.

Over past years the popularity of the uses of structures on the nanometer scale in the food sector is increasing, therefore, interest and activities in this research area have greatly focused. As nanobiotechnology steps forward, devices or material based on this technology become smaller and more sensitive. Its applicability in the areas of food packaging and food safety are well known. Additionally, promising results have been achieved in food preservation using nanomaterial where they might protect the food from moisture, lipids, gases, off-flavors, and odors. They offer excellent vehicle systems to deliver bioactive compounds to the target tissues. Although the advances in nanotechnology are paving new paths day by day, there still persist many challenges and opportunities to improve the current technology and also issues about the consequences of nanotechnology that must need to be addressed in order to alleviate consumer concerns. The transparency of safety issues and environmental impact should be the priority while dealing with the development of nanotechnology in food systems and therefore compulsory testing of nano foods is required before they are released to the market.

WET NANOTECHNOLOGY

Wet nanotechnology (also known as wet nanotech) involves working up to large masses from small ones.

Wet nanotechnology requires water in which the process occurs. The process also involves chemists and biologists trying to reach larger scales by putting together individual molecules. While Eric Drexler put forth the idea of nano-assemblers working dry, wet nanotech appears to be the likely first area in which something like a nano-assembler may achieve economic results. Pharmaceuticals and bioscience are central features of most nanotech start-ups. Richard A.L. Jones calls nanotechnology that steals bits of natural nanotechnology and puts them in a synthetic structure biokleptic nanotechnology. He calls building with synthetic materials according to nature's design principles biomimetic nanotechnology.

Using these guiding principles could lead to trillions of nanotech robots, that resemble bacteria in structural properties, entering a person's blood stream to do medical treatments.

Wet nanotechnology is an anticipated new sub-discipline of nanotech that is going to mostly be dominated by the different forms of wet engineering. The processes that will be used are going to take place in aqueous solutions and are very close to that of biotechnology manufacturing / bio-molecular manufacturing which is largely concerned with the production of biomolecules like proteins and DNA/RNA. There is some overlap of Biotechnology and Wet nanotechnology because living things are inherently bottom-up engineered and any exploitation of this by biotechnologists means they dabble in bottom-up engineering (though mostly at the level of producing macromolecules like proteins and nucleic acids from there monomer units. Wet nanotech, however, seeks to analyse living things and their components as engineering systems and aims to understand them completely to have complete control of the behavior of the system and to derive principles and methods that can be applied more broadly to bottom up manufacturing, to manipulate matter on the atomic and molecular scales and to creating machines or devices at the nanometer and microscopic scales. Biotech is mostly about exploiting living systems in anyway possible. Molecular Biology and related disciplines compare the mechanism of function of proteins in particular - and nucleic acids to a lesser extent - as like "molecular machines". In order for engineers to mimic these nanoscale machines in a way that they could be produced with some efficiency, they must look into bottom-up manufacturing. Bottom-up manufacturing deals with manipulating individual atoms during the manufacturing process, so that there is absolute control of their placement and interactions.

Then from the atomic scale, nanomachines could be made and even be designed to self-replicate themselves as long as they are designed in an environment with copious amount of the needed materials. Because individual atoms are being manipulated in the process, bottom-up manufacturing is often referred to as "atom by atom" manufacturing. If the manufacturing of nanomachines can be made more readily available through improved techniques, there could be a large economic and social impact. This would start with improvements in making microelectromechanical systems and then would allow for the creation of nanoscale biological sensors along with things that have not been thought of yet. This is because "wet" nanotech is only in the beginning of its life. Scientists and engineers alike feel that biomimetics is a great way to start looking at creating nanoscale machines. Humans have only had a few thousand years to try to learn about the mechanics of things at really small scales. However, nature has been working on perfecting the design and functionality of nanomachines for millions of years. This is why there are already nanomachines, such as ATP synthase, working in our bodies that have an unheard of 95% efficiency.

Wet vs. Dry

Wet nanotechnology is a form of wet engineering as opposed to dry engineering. There are different fields that deal with those two types of engineering. Biologists, from the point of view of nanotechnology, deal with wet engineering. They study processes that happen in life, and for the most part those processes take place in aqueous environments. Our bodies are made up mostly of water.

Electrical and mechanical engineers are on the other side of the line in dry engineering. They are involved with processes and manufacturing that does not occur in aqueous environments.

For the most part, wet engineering deals with "soft" materials that allow for flexibility which is vital at the nanoscale in biological manufacturing. Dry engineers mostly handle things with rigid structures and parts. These differences stem for the fact that the forces that the two types of engineers must deal with are very different. At a larger scale, most things are dominated by Newtonian physics. However, when one looks at the nanoscale, especially in biological matters, the dominating force is Brownian motion.

Because nanotechnology in the new age is going to most likely deal with both dry and wet in conjunction with each other, there is going to have to be a change in the way society looks at engineering and manufacturing. People will have to be not only well educated in engineering but also in biology because the integration of the two is how there will be the largest improvements in nanotechnology.

Brownian Motion as it Relates to Wet Nanotech

With the existence of natural nanomachines, "a complex precision microscopic-sized machine that fits the standard definition of a machine", such as ATP synthase and T4 bacteriophage, scientists and biologists know that they are capable of making similar types of machines at the same scale. However, nature has had a long time to perfect the building and creation of these nanomachines and humankind has only just begun to look into them with greater interest.

This interest may have been sparked because of the existence of nanomachines such as ATP synthase (adenosine triphosphate), which is the "second in importance only to DNA". ATP is the main energy converter that our bodies contain and without it, life as we know it would not be able to flourish or even survive.

What does Brownian Motion have to do with Complex Nanomachines?

Brownian motion is a random, constantly fluctuating force that acts on a body in environments that are at a microscale. This force is one that mechanical engineers and physicists are not used to dealing with because, at the larger scale that humankind tends to think of things, this force is not one that needs to be taken into account. People think of gravity, inertia, and other physics based forces that act on us all the time, however at the nanoscale those forces are mostly "negligible".

In order for nanomachines to be recreated by man, either there will need to be discoveries that allow us to understand how to "exploit" Brownian motion as nature does or find a way to work around it by using materials that are rigid enough to stand up to these forces. The way that nature has been able to exploit Brownian motion is through self-assembly. This force pushes and pulls all of the proteins and amino acids around in our bodies and sticks them together in all sorts of combinations. The combinations that do not work separate and continue with their random attachment however, the combinations that do work produce things like ATP synthase. Through this process nature has been able to make a nanomachine that is 95% efficient, which is a feat that man has not been able to accomplish yet. This is all because nature does not try to work around the forces; it uses them at its advantage.

Growing cells in culture to take advantage of their internal chemical synthesis machinery can be considered a form of nanotechnology but this machinery has also been manipulated outside of living cells.

DNA NANOTECHNOLOGY

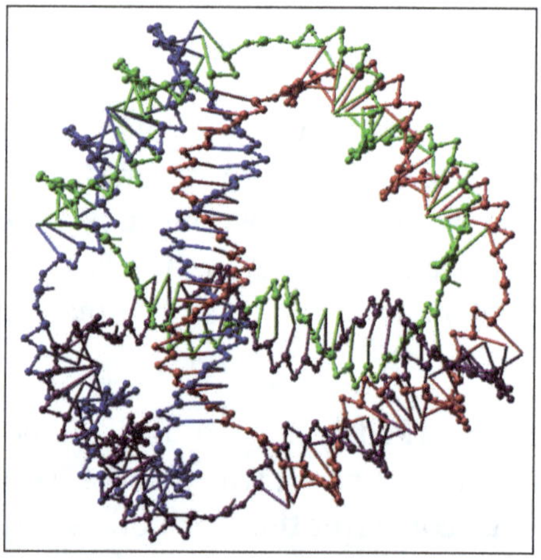

DNA nanotechnology involves forming artificial, designed nanostructures out of nucleic acids, such as this DNA tetrahedron. Each edge of the tetrahedron is a 20 base pair DNA double helix, and each vertex is a three-arm junction. The 4 DNA strands that form the 4 tetrahedral faces are color-coded.

DNA nanotechnology is the design and manufacture of artificial nucleic acid structures for technological uses. In this field, nucleic acids are used as non-biological engineering materials for nanotechnology rather than as the carriers of genetic information in living cells. Researchers in the field have created static structures such as two- and three-dimensional crystal lattices, nanotubes, polyhedra, and arbitrary shapes, and functional devices such as molecular machines and DNA computers. The field is beginning to be used as a tool to solve basic science problems in structural biology and biophysics, including applications in X-ray crystallography and nuclear magnetic resonance spectroscopy of proteins to determine structures. Potential applications in molecular scale electronics and nanomedicine are also being investigated.

The conceptual foundation for DNA nanotechnology was first laid out by Nadrian Seeman in the early 1980s, and the field began to attract widespread interest in the mid-2000s. This use of nucleic acids is enabled by their strict base pairing rules, which cause only portions of strands with complementary base sequences to bind together to form strong, rigid double helix structures. This allows for the rational design of base sequences that will selectively assemble to form complex target structures with precisely controlled nanoscale features. Several assembly methods are used to make these structures, including tile-based structures that assemble from smaller structures, folding structures using the DNA origami method, and dynamically reconfigurable structures using strand displacement methods. The field's name specifically references DNA, but the same principles have been used with other types of nucleic acids as well, leading to the occasional use of the alternative name nucleic acid nanotechnology.

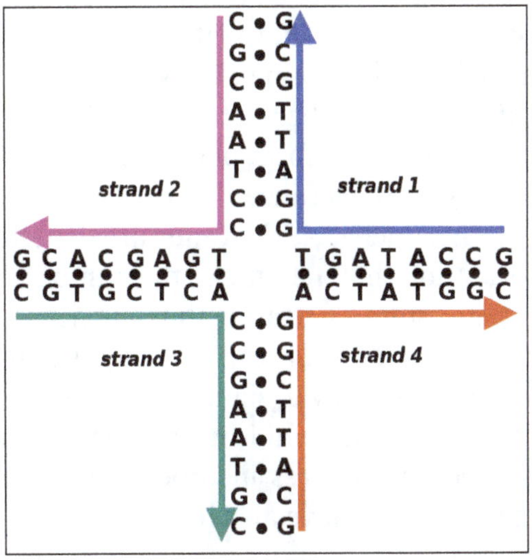

These four strands associate into a DNA four-arm junction because this structure maximizes the number of correct base pairs, with A matched to T and C matched to G.

This double-crossover (DX) supramolecular complex consists of five DNA single strands that form two double-helical domains, on the top and the bottom in this image. There are two crossover points where the strands cross from one domain into the other.

Properties of Nucleic Acids

Nanotechnology is often defined as the study of materials and devices with features on a scale below 100 nanometers. DNA nanotechnology, specifically, is an example of bottom-up molecular self-assembly, in which molecular components spontaneously organize into stable structures; the particular form of these structures is induced by the physical and chemical properties of the components selected by the designers. In DNA nanotechnology, the component materials are strands of nucleic acids such as DNA; these strands are often synthetic and are almost always used outside the context of a living cell. DNA is well-suited to nanoscale construction because the binding between two nucleic acid strands depends on simple base pairing rules which are well understood, and form the specific nanoscale structure of the nucleic acid double helix. These qualities make the assembly of nucleic acid structures easy to control through nucleic acid design. This property is absent in other materials used in nanotechnology, including proteins, for which protein design is very difficult, and nanoparticles, which lack the capability for specific assembly on their own.

The structure of a nucleic acid molecule consists of a sequence of nucleotides distinguished by which nucleobase they contain. In DNA, the four bases present are adenine (A), cytosine (C), guanine (G), and thymine (T). Nucleic acids have the property that two molecules will only bind to each other to form a double helix if the two sequences are complementary, meaning that they form matching sequences of base pairs, with A only binding to T, and C only to G. Because the formation of correctly matched base pairs is energetically favorable, nucleic acid strands are expected in most cases to bind to each other in the conformation that maximizes the number of correctly paired bases. The sequences of bases in a system of strands thus determine the pattern of binding and the overall structure in an easily controllable way. In DNA nanotechnology, the base sequences of strands are rationally designed by researchers so that the base pairing interactions cause the strands to assemble in the desired conformation. While DNA is the dominant material used, structures incorporating other nucleic acids such as RNA and peptide nucleic acid (PNA) have also been constructed.

Subfields

DNA nanotechnology is sometimes divided into two overlapping subfields: structural DNA nanotechnology and dynamic DNA nanotechnology. Structural DNA nanotechnology, sometimes abbreviated as SDN, focuses on synthesizing and characterizing nucleic acid complexes and materials that assemble into a static, equilibrium end state. On the other hand, dynamic DNA nanotechnology focuses on complexes with useful non-equilibrium behavior such as the ability to reconfigure based on a chemical or physical stimulus. Some complexes, such as nucleic acid nanomechanical devices, combine features of both the structural and dynamic subfields.

The complexes constructed in structural DNA nanotechnology use topologically branched nucleic acid structures containing junctions. (In contrast, most biological DNA exists as an unbranched double helix.) One of the simplest branched structures is a four-arm junction that consists of four individual DNA strands, portions of which are complementary in a specific pattern. Unlike in natural Holliday junctions, each arm in the artificial immobile four-arm junction has a different base sequence, causing the junction point to be fixed at a certain position. Multiple junctions can be combined in the same complex, such as in the widely used double-crossover (DX) structural motif, which contains two parallel double helical domains with individual strands crossing between the domains at two crossover points. Each crossover point is, topologically, a four-arm junction, but is constrained to one orientation, in contrast to the flexible single four-arm junction, providing a rigidity that makes the DX motif suitable as a structural building block for larger DNA complexes.

Dynamic DNA nanotechnology uses a mechanism called toehold-mediated strand displacement to allow the nucleic acid complexes to reconfigure in response to the addition of a new nucleic acid strand. In this reaction, the incoming strand binds to a single-stranded toehold region of a double-stranded complex, and then displaces one of the strands bound in the original complex through a branch migration process. The overall effect is that one of the strands in the complex is replaced with another one. In addition, reconfigurable structures and devices can be made using functional nucleic acids such as deoxyribozymes and ribozymes, which can perform chemical reactions, and aptamers, which can bind to specific proteins or small molecules.

Structural DNA Nanotechnology

Structural DNA nanotechnology, sometimes abbreviated as SDN, focuses on synthesizing and characterizing nucleic acid complexes and materials where the assembly has a static, equilibrium endpoint. The nucleic acid double helix has a robust, defined three-dimensional geometry that makes it possible to predict and design the structures of more complicated nucleic acid complexes. Many such structures have been created, including two- and three-dimensional structures, and periodic, aperiodic, and discrete structures.

Extended Lattices

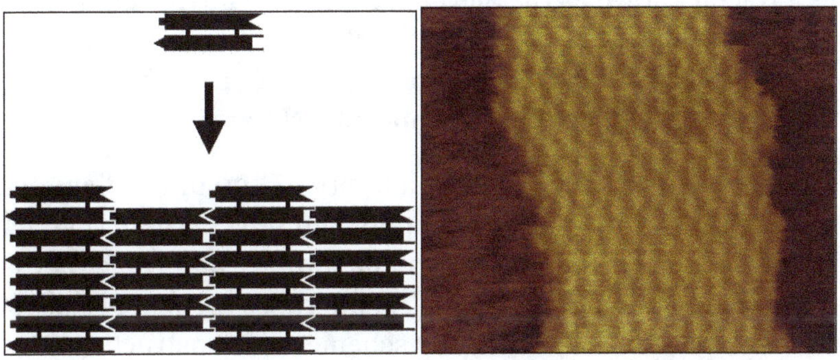

The assembly of a DX array. *Left*, schematic diagram. Each bar represents a double-helical domain of DNA, with the shapes representing complementary sticky ends. The DX complex at top will combine with other DX complexes into the two-dimensional array shown at bottom. *Right*, an atomic force microscopy image of the assembled array. The individual DX tiles are clearly visible within the assembled structure. The field is 150 nm across.

Small nucleic acid complexes can be equipped with sticky ends and combined into larger two-dimensional periodic lattices containing a specific tessellated pattern of the individual molecular tiles. The earliest example of this used double-crossover (DX) complexes as the basic tiles, each containing four sticky ends designed with sequences that caused the DX units to combine into periodic two-dimensional flat sheets that are essentially rigid two-dimensional crystals of DNA. Two-dimensional arrays have been made from other motifs as well, including the Holliday junction rhombus lattice, and various DX-based arrays making use of a double-cohesion scheme. The top two images at right show examples of tile-based periodic lattices.

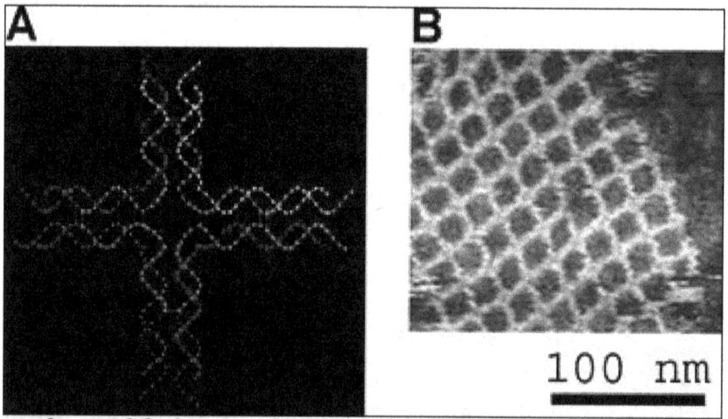

Left, a model of a DNA tile used to make another two-dimensional periodic lattice. *Right*, an atomic force micrograph of the assembled lattice.

Two-dimensional arrays can be made to exhibit aperiodic structures whose assembly implements a specific algorithm, exhibiting one form of DNA computing. The DX tiles can have their sticky end sequences chosen so that they act as Wang tiles, allowing them to perform computation. A DX array whose assembly encodes an XOR operation has been demonstrated; this allows the DNA array to implement a cellular automaton that generates a fractal known as the Sierpinski gasket. The third image at right shows this type of array. Another system has the function of a binary counter, displaying a representation of increasing binary numbers as it grows. These results show that computation can be incorporated into the assembly of DNA arrays.

DX arrays have been made to form hollow nanotubes 4–20 nm in diameter, essentially two-dimensional lattices which curve back upon themselves. These DNA nanotubes are somewhat similar in size and shape to carbon nanotubes, and while they lack the electrical conductance of carbon nanotubes, DNA nanotubes are more easily modified and connected to other structures. One of many schemes for constructing DNA nanotubes uses a lattice of curved DX tiles that curls around itself and closes into a tube. In an alternative method that allows the circumference to be specified in a simple, modular fashion using single-stranded tiles, the rigidity of the tube is an emergent property.

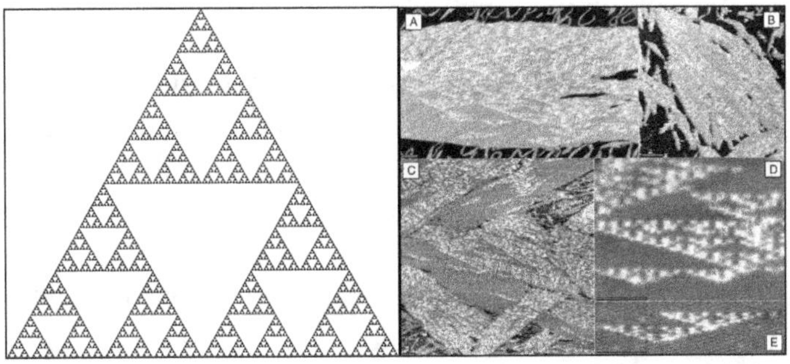

An example of an aperiodic two-dimensional lattice that assembles into a fractal pattern. *Left*, the Sierpinski gasket fractal. *Right*, DNA arrays that display a representation of the Sierpinski gasket on their surfaces.

Forming three-dimensional lattices of DNA was the earliest goal of DNA nanotechnology, but this proved to be one of the most difficult to realize. Success using a motif based on the concept of tensegrity, a balance between tension and compression forces, was finally reported in 2009.

Discrete Structures

Researchers have synthesized many three-dimensional DNA complexes that each have the connectivity of a polyhedron, such as a cube or octahedron, meaning that the DNA duplexes trace the edges of a polyhedron with a DNA junction at each vertex. The earliest demonstrations of DNA polyhedra were very work-intensive, requiring multiple ligations and solid-phase synthesis steps to create catenated polyhedra. Subsequent work yielded polyhedra whose synthesis was much easier. These include a DNA octahedron made from a long single strand designed to fold into the correct conformation, and a tetrahedron that can be produced from four DNA strands in one step.

Nanostructures of arbitrary, non-regular shapes are usually made using the DNA origami method. These structures consist of a long, natural virus strand as a "scaffold", which is made to fold into the desired shape by computationally designed short "staple" strands. This method has the advantages of being easy to design, as the base sequence is predetermined by the scaffold strand sequence, and not requiring high strand purity and accurate stoichiometry, as most other DNA nanotechnology methods do. DNA origami was first demonstrated for two-dimensional shapes, such as a smiley face, a coarse map of the Western Hemisphere, and the Mona Lisa painting. Solid three-dimensional structures can be made by using parallel DNA helices arranged in a honeycomb pattern, and structures with two-dimensional faces can be made to fold into a hollow overall three-dimensional shape, akin to a cardboard box. These can be programmed to open and reveal or release a molecular cargo in response to a stimulus, making them potentially useful as programmable molecular cages.

Templated Assembly

Nucleic acid structures can be made to incorporate molecules other than nucleic acids, sometimes called heteroelements, including proteins, metallic nanoparticles, quantum dots, and fullerenes. This allows the construction of materials and devices with a range of functionalities much greater than is possible with nucleic acids alone. The goal is to use the self-assembly of the nucleic acid structures to template the assembly of the nanoparticles hosted on them, controlling their position and in some cases orientation. Many of these schemes use a covalent attachment scheme, using oligonucleotides with amide or thiol functional groups as a chemical handle to bind the heteroelements. This covalent binding scheme has been used to arrange gold nanoparticles on a DX-based array, and to arrange streptavidin protein molecules into specific patterns on a DX array. A non-covalent hosting scheme using Dervan polyamides on a DX array was used to arrange streptavidin proteins in a specific pattern on a DX array. Carbon nanotubes have been hosted on DNA arrays in a pattern allowing the assembly to act as a molecular electronic device, a carbon nanotube field-effect transistor. In addition, there are nucleic acid metallization methods, in which the nucleic acid is replaced by a metal which assumes the general shape of the original nucleic acid structure, and schemes for using nucleic acid nanostructures as lithography masks, transferring their pattern into a solid surface.

Dynamic DNA Nanotechnology

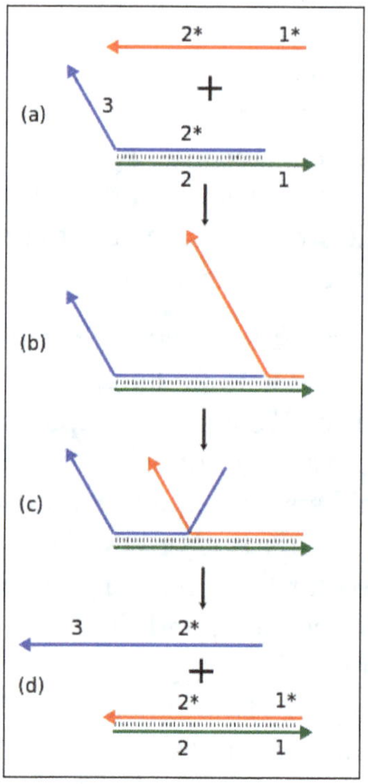

Dynamic DNA nanotechnology often makes use of toehold-mediated strand displacement reactions. In this example, the red strand binds to the single stranded toehold region on the green strand (region 1), and then in a branch migration process across region 2, the blue strand is displaced and freed from the complex. Reactions like these are used to dynamically reconfigure or assemble nucleic acid nanostructures. In addition, the red and blue strands can be used as signals in a molecular logic gate.

Dynamic DNA nanotechnology focuses on forming nucleic acid systems with designed dynamic functionalities related to their overall structures, such as computation and mechanical motion. There is some overlap between structural and dynamic DNA nanotechnology, as structures can be formed through annealing and then reconfigured dynamically, or can be made to form dynamically in the first place.

Nanomechanical Devices

DNA complexes have been made that change their conformation upon some stimulus, making them one form of nanorobotics. These structures are initially formed in the same way as the static structures made in structural DNA nanotechnology, but are designed so that dynamic reconfiguration is possible after the initial assembly. The earliest such device made use of the transition between the B-DNA and Z-DNA forms to respond to a change in buffer conditions by undergoing a twisting motion. This reliance on buffer conditions caused all devices to change state at the same time. Subsequent systems could change states based upon the presence of control strands, allowing multiple devices to be independently operated in solution. Some examples of such systems are a "molecular tweezers" design that has an open and a closed state, a device that could switch from a paranemic-crossover (PX) conformation to a double-junction (JX2) conformation, undergoing rotational motion in the process, and a two-dimensional array that could dynamically expand and contract in response to control strands. Structures have also been made that dynamically open or close, potentially acting as a molecular cage to release or reveal a functional cargo upon opening.

DNA walkers are a class of nucleic acid nanomachines that exhibit directional motion along a linear track. A large number of schemes have been demonstrated. One strategy is to control the motion of the walker along the track using control strands that need to be manually added in sequence. Another approach is to make use of restriction enzymes or deoxyribozymes to cleave the strands and cause the walker to move forward, which has the advantage of running autonomously. A later system could walk upon a two-dimensional surface rather than a linear track, and demonstrated the ability to selectively pick up and move molecular cargo. Additionally, a linear walker has been demonstrated that performs DNA-templated synthesis as the walker advances along the track, allowing autonomous multistep chemical synthesis directed by the walker. The synthetic DNA walkers' function is similar to that of the proteins dynein and kinesin.

Strand Displacement Cascades

Cascades of strand displacement reactions can be used for either computational or structural purposes. An individual strand displacement reaction involves revealing a new sequence in response to the presence of some initiator strand. Many such reactions can be linked into a cascade where the newly revealed output sequence of one reaction can initiate another strand displacement reaction elsewhere. This in turn allows for the construction of chemical reaction networks with many components, exhibiting complex computational and information processing abilities. These cascades are made energetically favorable through the formation of new base pairs, and the entropy gain from disassembly reactions. Strand displacement cascades allow isothermal operation of the assembly or computational process, in contrast to traditional nucleic acid assembly's requirement for a thermal annealing step, where the temperature is raised and then slowly lowered to ensure proper formation of the desired structure. They can also support catalytic function of the initiator species, where less than one equivalent of the initiator can cause the reaction to go to completion.

Strand displacement complexes can be used to make molecular logic gates capable of complex computation. Unlike traditional electronic computers, which use electric current as inputs and outputs, molecular computers use the concentrations of specific chemical species as signals. In the case of nucleic acid strand displacement circuits, the signal is the presence of nucleic acid strands that are released or consumed by binding and unbinding events to other strands in displacement complexes. This approach has been used to make logic gates such as and, or, and not gates. More recently, a four-bit circuit was demonstrated that can compute the square root of the integers 0–15, using a system of gates containing 130 DNA strands.

Another use of strand displacement cascades is to make dynamically assembled structures. These use a hairpin structure for the reactants, so that when the input strand binds, the newly revealed sequence is on the same molecule rather than disassembling. This allows new opened hairpins to be added to a growing complex. This approach has been used to make simple structures such as three- and four-arm junctions and dendrimers.

Applications

DNA nanotechnology provides one of the few ways to form designed, complex structures with precise control over nanoscale features. The field is beginning to see application to solve basic science problems in structural biology and biophysics. The earliest such application envisaged for the field, and one still in development, is in crystallography, where molecules that are difficult to crystallize in isolation could be arranged within a three-dimensional nucleic acid lattice, allowing determination of their structure. Another application is the use of DNA origami rods to replace liquid crystals in residual dipolar coupling experiments in protein NMR spectroscopy; using DNA origami is advantageous because, unlike liquid crystals, they are tolerant of the detergents needed to suspend membrane proteins in solution. DNA walkers have been used as nanoscale assembly lines to move nanoparticles and direct chemical synthesis. Further, DNA origami structures have aided in the biophysical studies of enzyme function and protein folding.

DNA nanotechnology is moving toward potential real-world applications. The ability of nucleic acid arrays to arrange other molecules indicates its potential applications in molecular scale electronics. The assembly of a nucleic acid structure could be used to template the assembly of a molecular electronic element such as molecular wires, providing a method for nanometer-scale control of the placement and overall architecture of the device analogous to a molecular breadboard. DNA nanotechnology has been compared to the concept of programmable matter because of the coupling of computation to its material properties.

In a study conducted by a group of scientists from iNANO and CDNA centers in Aarhus University, researchers were able to construct a small multi-switchable 3D DNA Box Origami. The proposed nanoparticle was characterized by atomic force microscopy (AFM), transmission electron microscopy (TEM) and Förster resonance energy transfer (FRET). The constructed box was shown to have a unique reclosing mechanism, which enabled it to repeatedly open and close in response to a unique set of DNA or RNA keys. The authors proposed that this "DNA device can potentially be used for a broad range of applications such as controlling the function of single molecules, controlled drug delivery, and molecular computing."

There are potential applications for DNA nanotechnology in nanomedicine, making use of its ability to perform computation in a biocompatible format to make "smart drugs" for targeted drug delivery, as well as for diagnostic applications. One such system being investigated uses a hollow DNA box containing proteins that induce apoptosis, or cell death, that will only open when in proximity to a cancer cell. There has additionally been interest in expressing these artificial structures in engineered living bacterial cells, most likely using the transcribed RNA for the assembly, although it is unknown whether these complex structures are able to efficiently fold or assemble in the cell's cytoplasm. If successful, this could enable directed evolution of nucleic acid nanostructures. Scientists at Oxford University reported the self-assembly of four short strands of synthetic DNA into a cage which can enter cells and survive for at least 48 hours. The fluorescently labeled DNA tetrahedra were found to remain intact in the laboratory cultured human kidney cells despite the attack by cellular enzymes after two days. This experiment showed the potential of drug delivery inside the living cells using the DNA 'cage'. A DNA tetrahedron was used to deliver RNA Interference (RNAi) in a mouse model, reported a team of researchers in MIT. Delivery of the interfering RNA for treatment has showed some success using polymer or lipid, but there are limits of safety and imprecise targeting, in addition to short shelf life in the blood stream. The DNA nanostructure created by the team consists of six strands of DNA to form a tetrahedron, with one strand of RNA affixed to each of the six edges. The tetrahedron is further equipped with targeting protein, three folate molecules, which lead the DNA nanoparticles to the abundant folate receptors found on some tumors. The result showed that the gene expression targeted by the RNAi, luciferase, dropped by more than half. This study shows promise in using DNA nanotechnology as an effective tool to deliver treatment using the emerging RNA Interference technology. The DNA tetrahedron was also used in an effort to overcome the phenomena multidrug resistance. Doxorubicin (DOX) was conjugated with the tetrahedron and was loaded into MCF-7 breast cancer cells that contained the P-glycoprotein drug efflux pump. The results of the experiment showed the DOX was not being pumped out and apoptosis of the cancer cells was achieved. The tetrahedron without DOX was loaded into cells to test its biocompatibility, and the structure showed no cytotoxicity itself. The DNA tetrahedron was also used as barcode for profiling the subcellular expression and distribution of proteins in cells for diagnostic purposes. The tetrahedral-nanostructured showed enhanced signal due to higher labeling efficiency and stability.

Applications for DNA nanotechnology in nanomedicine also focus on mimicking the structure and function of naturally occurring membrane proteins with designed DNA nanostructures. In 2012, Langecker introduced a pore-shaped DNA origami structure that can self-insert into lipid membranes via hydrophobic cholesterol modifications and induce ionic currents across the membrane. This first demonstration of a synthetic DNA ion channel was followed by a variety of pore-inducing designs ranging from a single DNA duplex, to small tile-based structures, and large DNA origami transmembrane porins. Similar to naturally occurring protein ion channels, this ensemble of synthetic DNA-made counterparts thereby spans multiple orders of magnitude in conductance. The study of the membrane-inserting single DNA duplex showed that current must also flow on the DNA-lipid interface as no central channel lumen is present in the design that lets ions pass across the lipid bilayer. This indicated that the DNA-induced lipid pore has a toroidal shape, rather than cylindrical, as lipid headgroups reorient to face towards the membrane-inserted part of the DNA. Researchers from the University of Cambridge and the University of Illinois at Urbana-Champaign then demonstrated that such a DNA-induced toroidal pore can facilitate rapid lipid flip-flop between the lipid bilayer leaflets. Utilizing this effect, they designed a synthetic DNA-built enzyme that flips lipids in biological membranes orders of magnitudes faster than naturally occurring proteins called scramblases. This development highlights the potential of synthetic DNA nanostructures for personalized drugs and therapeutics.

Design

DNA nanostructures must be rationally designed so that individual nucleic acid strands will assemble into the desired structures. This process usually begins with specification of a desired target structure or function. Then, the overall secondary structure of the target complex is determined, specifying the arrangement of nucleic acid strands within the structure, and which portions of those strands should be bound to each other. The last step is the primary structure design, which is the specification of the actual base sequences of each nucleic acid strand.

Structural Design

The first step in designing a nucleic acid nanostructure is to decide how a given structure should be represented by a specific arrangement of nucleic acid strands. This design step determines the secondary structure, or the positions of the base pairs that hold the individual strands together in the desired shape. Several approaches have been demonstrated:

- Tile-based structures: This approach breaks the target structure into smaller units with strong binding between the strands contained in each unit, and weaker interactions between the units. It is often used to make periodic lattices, but can also be used to implement algorithmic self-assembly, making them a platform for DNA computing. This was the dominant design strategy used from the mid-1990s until the mid-2000s, when the DNA origami methodology was developed.
- Folding structures: An alternative to the tile-based approach, folding approaches make the nanostructure from one long strand, which can either have a designed sequence that folds due to its interactions with itself, or it can be folded into the desired shape by using shorter, "staple" strands. This latter method is called DNA origami, which allows forming nanoscale two- and three-dimensional shapes.
- Dynamic assembly: This approach directly controls the kinetics of DNA self-assembly, specifying all of the intermediate steps in the reaction mechanism in addition to the final product. This is done using starting materials which adopt a hairpin structure; these then assemble into the final conformation in a cascade reaction, in a specific order. This approach has the advantage of proceeding isothermally, at a constant temperature. This is in contrast to the thermodynamic approaches, which require a thermal annealing step where a temperature change is required to trigger the assembly and favor proper formation of the desired structure.

Sequence Design

After any of the above approaches are used to design the secondary structure of a target complex, an actual sequence of nucleotides that will form into the desired structure must be devised. Nucleic acid design is the process of assigning a specific nucleic acid base sequence to each of a structure's constituent strands so that they will associate into a desired conformation. Most methods have the goal of designing sequences so that the target structure has the lowest energy, and is thus the most thermodynamically favorable, while incorrectly assembled structures have higher energies and are thus disfavored. This is done either through simple, faster heuristic methods such as sequence symmetry minimization, or by using a full nearest-neighbor thermodynamic model, which is more accurate but slower and more computationally intensive.

Geometric models are used to examine tertiary structure of the nanostructures and to ensure that the complexes are not overly strained.

Nucleic acid design has similar goals to protein design. In both, the sequence of monomers is designed to favor the desired target structure and to disfavor other structures. Nucleic acid design has the advantage of being much computationally easier than protein design, because the simple base pairing rules are sufficient to predict a structure's energetic favorability, and detailed information about the overall three-dimensional folding of the structure is not required. This allows the use of simple heuristic methods that yield experimentally robust designs. Nucleic acid structures are less versatile than proteins in their function because of proteins' increased ability to fold into complex structures, and the limited chemical diversity of the four nucleotides as compared to the twenty proteinogenic amino acids.

Materials and Methods

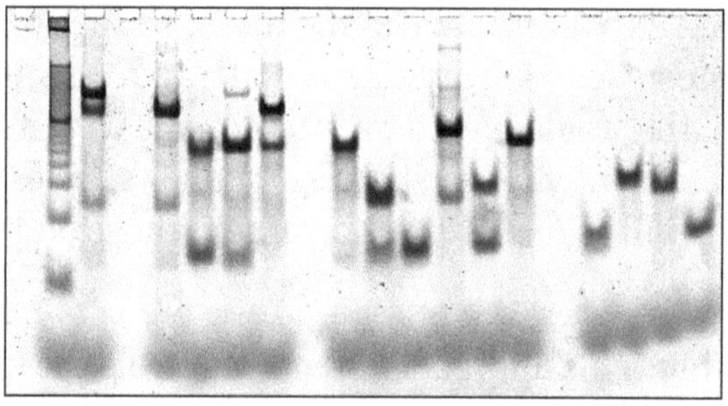

Gel electrophoresis methods, such as this formation assay on a DX complex, are used to ascertain whether the desired structures are forming properly. Each vertical lane contains a series of bands, where each band is characteristic of a particular reaction intermediate.

The sequences of the DNA strands making up a target structure are designed computationally, using molecular modeling and thermodynamic modeling software. The nucleic acids themselves are then synthesized using standard oligonucleotide synthesis methods, usually automated in an oligonucleotide synthesizer, and strands of custom sequences are commercially available. Strands can be purified by denaturing gel electrophoresis if needed, and precise concentrations determined via any of several nucleic acid quantitation methods using ultraviolet absorbance spectroscopy.

The fully formed target structures can be verified using native gel electrophoresis, which gives size and shape information for the nucleic acid complexes. An electrophoretic mobility shift assay can assess whether a structure incorporates all desired strands. Fluorescent labeling and Förster resonance energy transfer (FRET) are sometimes used to characterize the structure of the complexes.

Nucleic acid structures can be directly imaged by atomic force microscopy, which is well suited to extended two-dimensional structures, but less useful for discrete three-dimensional structures because of the microscope tip's interaction with the fragile nucleic acid structure; transmission electron microscopy and cryo-electron microscopy are often used in this case. Extended three-dimensional lattices are analyzed by X-ray crystallography.

NANOROBOTICS

Nanorobotics is an emerging technology field creating machines or robots whose components are at or near the scale of a nanometer (10^{-9} meters). More specifically, nanorobotics (as opposed to microrobotics) refers to the nanotechnology engineering discipline of designing and building nanorobots, with devices ranging in size from 0.1–10 micrometres and constructed of nanoscale or molecular components. The terms *nanobot, nanoid, nanite, nanomachine*, or *nanomite* have also been used to describe such devices currently under research and development.

Nanomachines are largely in the research and development phase, but some primitive molecular machines and nanomotors have been tested. An example is a sensor having a switch approximately 1.5 nanometers across, able to count specific molecules in a chemical sample. The first useful applications of nanomachines may be in nanomedicine. For example, biological machines could be used to identify and destroy cancer cells. Another potential application is the detection of toxic chemicals, and the measurement of their concentrations, in the environment. Rice University has demonstrated a single-molecule car developed by a chemical process and including Buckminsterfullerenes (buckyballs) for wheels. It is actuated by controlling the environmental temperature and by positioning a scanning tunneling microscope tip.

Another definition is a robot that allows precise interactions with nanoscale objects, or can manipulate with nanoscale resolution. Such devices are more related to microscopy or scanning probe microscopy, instead of the description of nanorobots as molecular machines. Using the microscopy definition, even a large apparatus such as an atomic force microscope can be considered a nanorobotic instrument when configured to perform nanomanipulation. For this viewpoint, macroscale robots or microrobots that can move with nanoscale precision can also be considered nanorobots.

Nanorobotics Theory

According to Richard Feynman, it was his former graduate student and collaborator Albert Hibbs who originally suggested to him (circa 1959) the idea of a *medical* use for Feynman's theoretical micromachines. Hibbs suggested that certain repair machines might one day be reduced in size to the point that it would, in theory, be possible to (as

Feynman put it) "*swallow the surgeon*". The idea was incorporated into Feynman's 1959 essay *There's Plenty of Room at the Bottom.*

Since nanorobots would be microscopic in size, it would probably be necessary for very large numbers of them to work together to perform microscopic and macroscopic tasks. These nanorobot swarms, both those unable to replicate (as in utility fog) and those able to replicate unconstrainedly in the natural environment (as in grey goo and synthetic biology), are found in many science fiction stories, such as the Borg nanoprobes in *Star Trek* and *The Outer Limits* episode "The New Breed". Some proponents of nanorobotics, in reaction to the grey goo scenarios that they earlier helped to propagate, hold the view that nanorobots able to replicate outside of a restricted factory environment do not form a necessary part of a purported productive nanotechnology, and that the process of self-replication, were it ever to be developed, could be made inherently safe. They further assert that their current plans for developing and using molecular manufacturing do not in fact include free-foraging replicators.

A detailed theoretical discussion of nanorobotics, including specific design issues such as sensing, power communication, navigation, manipulation, locomotion, and onboard computation, has been presented in the medical context of nanomedicine by Robert Freitas. Some of these discussions remain at the level of unbuildable generality and do not approach the level of detailed engineering.

Legal and Ethical Implications

Open Technology

A document with a proposal on nanobiotech development using open design technology methods, as in open-source hardware and open-source software, has been addressed to the United Nations General Assembly. According to the document sent to the United Nations, in the same way that open source has in recent years accelerated the development of computer systems, a similar approach should benefit the society at large and accelerate nanorobotics development. The use of nanobiotechnology should be established as a human heritage for the coming generations, and developed as an open technology based on ethical practices for peaceful purposes. Open technology is stated as a fundamental key for such an aim.

Nanorobot Race

In the same ways that technology research and development drove the space race and nuclear arms race, a race for nanorobots is occurring. There is plenty of ground allowing nanorobots to be included among the emerging technologies. Some of the reasons are that large corporations, such as General Electric, Hewlett-Packard, Synopsys, Northrop Grumman and Siemens have been recently working in the development and research of nanorobots; surgeons are getting involved and starting to propose ways to apply nanorobots for common medical procedures; universities and research institutes were granted funds by government agencies exceeding $2 billion towards research developing nanodevices for medicine; bankers are also strategically investing with the intent to acquire beforehand rights and royalties on future nanorobots commercialisation. Some aspects of nanorobot litigation and related issues linked to monopoly have already arisen. A large number of patents has been granted recently on nanorobots, done mostly for patent agents, companies specialized solely on building patent portfolios, and lawyers. After a long series of patents and eventually litigations, emerging fields of technology tend to become a monopoly, which normally is dominated by large corporations.

Manufacturing Approaches

Manufacturing nanomachines assembled from molecular components is a very challenging task. Because of the level of difficulty, many engineers and scientists continue working cooperatively across multidisciplinary approaches to achieve breakthroughs in this new area of development. Thus, it is quite understandable the importance of the following distinct techniques currently applied towards manufacturing nanorobots:

Biochip

The joint use of nanoelectronics, photolithography, and new biomaterials provides a possible approach to manufacturing nanorobots for common medical uses, such as surgical instrumentation, diagnosis, and drug delivery. This method for manufacturing on nanotechnology scale is in use in the electronics industry since 2008. So, practical nanorobots should be integrated as nanoelectronics devices, which will allow tele-operation and advanced capabilities for medical instrumentation.

Nubots

A nucleic acid robot (nubot) is an organic molecular machine at the nanoscale. DNA structure can provide means to assemble 2D and 3D nanomechanical devices. DNA based machines can be activated using small molecules, proteins and other molecules of DNA. Biological circuit gates based on DNA materials have been engineered as molecular machines to allow in-vitro drug delivery for targeted health problems. Such material based systems would work most closely to smart biomaterial drug system delivery, while not allowing precise in vivo teleoperation of such engineered prototypes.

Surface-bound Systems

Several reports have demonstrated the attachment of synthetic molecular motors to surfaces. These primitive nanomachines have been shown to undergo machine-like motions when confined to the surface of a macroscopic material. The surface anchored motors could potentially be used to move and position nanoscale materials on a surface in the manner of a conveyor belt.

Positional Nanoassembly

Nanofactory Collaboration, founded by Robert Freitas and Ralph Merkle in 2000 and involving 23 researchers from 10 organizations and 4 countries, focuses on developing a practical research agenda specifically aimed at developing positionally-controlled diamond mechanosynthesis and a diamondoid nanofactory that would have the capability of building diamondoid medical nanorobots.

Biohybrids

The emerging field of bio-hybrid systems combines biological and synthetic structural elements for biomedical or robotic applications. The constituting elements of bio-nanoelectromechanical systems (BioNEMS) are of nanoscale size, for example DNA, proteins or nanostructured mechanical parts. Thiol-ene ebeam resist allow the direct writing of nanoscale features, followed by the functionalization of the natively reactive resist surface with biomolecules. Other approaches use a biodegradable material attached to magnetic particles that allow them to be guided around the body.

Bacteria-based

This approach proposes the use of biological microorganisms, like the bacterium *Escherichia coli* and *Salmonella typhimurium*. Thus the model uses a flagellum for propulsion purposes. Electromagnetic fields normally control the motion of this kind of biological integrated device. Chemists at the University of Nebraska have created a humidity gauge by fusing a bacterium to a silicone computer chip.

Virus-based

Retroviruses can be retrained to attach to cells and replace DNA. They go through a process called reverse transcription to deliver genetic packaging in a vector. Usually, these devices are Pol – Gag genes of the virus for the Capsid and Delivery system. This process is called retroviral gene therapy, having the ability to re-engineer cellular DNA by usage of viral vectors. This approach has appeared in the form of retroviral, adenoviral, and lentiviral gene delivery systems. These gene therapy vectors have been used in cats to send genes into the genetically modified organism (GMO), causing it to display the trait.

3D Printing

3D printing is the process by which a three-dimensional structure is built through the various processes of additive manufacturing. Nanoscale 3D printing involves many of the same process, incorporated at a much smaller scale. To print a structure in the 5-400 µm scale, the precision of the 3D printing machine is improved greatly. A two-steps process of 3D printing, using a 3D printing and laser etched plates method was incorporated as an improvement technique. To be more precise at a nanoscale, the

3D printing process uses a laser etching machine, which etches into each plate the details needed for the segment of nanorobot. The plate is then transferred to the 3D printer, which fills the etched regions with the desired nanoparticle. The 3D printing process is repeated until the nanorobot is built from the bottom up. This 3D printing process has many benefits. First, it increases the overall accuracy of the printing process. Second, it has the potential to create functional segments of a nanorobot. The 3D printer uses a liquid resin, which is hardened at precisely the correct spots by a focused laser beam. The focal point of the laser beam is guided through the resin by movable mirrors and leaves behind a hardened line of solid polymer, just a few hundred nanometers wide. This fine resolution enables the creation of intricately structured sculptures as tiny as a grain of sand. This process takes place by using photoactive resins, which are hardened by the laser at an extremely small scale to create the structure. This process is quick by nanoscale 3D printing standards. Ultra-small features can be made with the 3D micro-fabrication technique used in multiphoton photopolymerisation. This approach uses a focused laser to trace the desired 3D object into a block of gel. Due to the nonlinear nature of photo excitation, the gel is cured to a solid only in the places where the laser was focused while the remaining gel is then washed away. Feature sizes of under 100 nm are easily produced, as well as complex structures with moving and interlocked parts.

Potential Uses

Nanomedicine

Potential uses for nanorobotics in medicine include early diagnosis and targeted drug-delivery for cancer, biomedical instrumentation, surgery, pharmacokinetics, monitoring of diabetes and health care.

In such plans, future medical nanotechnology is expected to employ nanorobots injected into the patient to perform work at a cellular level. Such nanorobots intended for use in medicine should be non-replicating, as replication would needlessly increase device complexity, reduce reliability, and interfere with the medical mission.

Nanotechnology provides a wide range of new technologies for developing customized means to optimize the delivery of pharmaceutical drugs. Today, harmful side effects of treatments such as chemotherapy are commonly a result of drug delivery methods that don't pinpoint their intended target cells accurately. Researchers at Harvard and MIT, however, have been able to attach special RNA strands, measuring nearly 10 nm in diameter, to nanoparticles, filling them with a chemotherapy drug. These RNA strands are attracted to cancer cells. When the nanoparticle encounters a cancer cell, it adheres to it, and releases the drug into the cancer cell. This directed method of drug delivery has great potential for treating cancer patients while avoiding negative effects (commonly associated with improper drug delivery). The first demonstration of nanomotors operating in living organisms was carried out in 2014 at University of

California, San Diego. MRI-guided nanocapsules are one potential precursor to nanorobots.

Another useful application of nanorobots is assisting in the repair of tissue cells alongside white blood cells. Recruiting inflammatory cells or white blood cells (which include neutrophil granulocytes, lymphocytes, monocytes, and mast cells) to the affected area is the first response of tissues to injury. Because of their small size, nanorobots could attach themselves to the surface of recruited white cells, to squeeze their way out through the walls of blood vessels and arrive at the injury site, where they can assist in the tissue repair process. Certain substances could possibly be used to accelerate the recovery.

The science behind this mechanism is quite complex. Passage of cells across the blood endothelium, a process known as transmigration, is a mechanism involving engagement of cell surface receptors to adhesion molecules, active force exertion and dilation of the vessel walls and physical deformation of the migrating cells. By attaching themselves to migrating inflammatory cells, the robots can in effect "hitch a ride" across the blood vessels, bypassing the need for a complex transmigration mechanism of their own.

As of 2016, in the United States, Food and Drug Administration (FDA) regulates nanotechnology on the basis of size.

Soutik Betal, during his doctoral research at the University of Texas, San Antonio developed nanocomposite particles that are controlled remotely by an electromagnetic field. This series of nanorobots that are now enlisted in the Guinness World Records, can be used to interact with the biological cells. Scientists suggest that this technology can be used for the treatment of cancer.

NANOMEDICINE

Nanomedicine is the medical application of nanotechnology. Nanomedicine ranges from the medical applications of nanomaterials and biological devices, to nanoelectronic biosensors, and even possible future applications of molecular nanotechnology such as biological machines. Current problems for nanomedicine involve understanding the issues related to toxicity and environmental impact of nanoscale materials (materials whose structure is on the scale of nanometers, i.e. billionths of a meter).

Functionalities can be added to nanomaterials by interfacing them with biological molecules or structures. The size of nanomaterials is similar to that of most biological molecules and structures; therefore, nanomaterials can be useful for both in vivo and in vitro biomedical research and applications. Thus far, the integration of nanomaterials with biology has led to the development of diagnostic devices, contrast agents, analytical tools, physical therapy applications, and drug delivery vehicles.

Nanomedicine seeks to deliver a valuable set of research tools and clinically useful devices in the near future. The National Nanotechnology Initiative expects new commercial applications in the pharmaceutical industry that may include advanced drug delivery systems, new therapies, and in vivo imaging. Nanomedicine research is receiving funding from the US National Institutes of Health Common Fund program, supporting four nanomedicine development centers.

Nanomedicine sales reached $16 billion in 2015, with a minimum of $3.8 billion in nanotechnology R&D being invested every year. Global funding for emerging nanotechnology increased by 45% per year in recent years, with product sales exceeding $1 trillion in 2013. As the nanomedicine industry continues to grow, it is expected to have a significant impact on the economy.

Drug Delivery

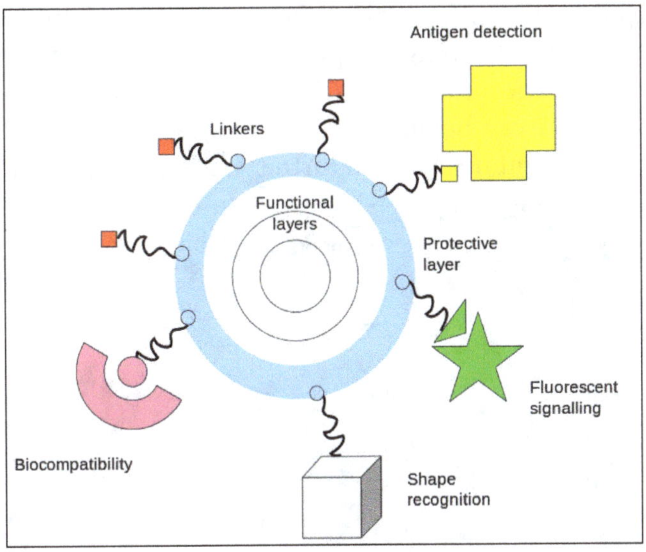

Nanotechnology has provided the possibility of delivering drugs to specific cells using nanoparticles. The overall drug consumption and side-effects may be lowered significantly by depositing the active agent in the morbid region only and in no higher dose than needed. Targeted drug delivery is intended to reduce the side effects of drugs with concomitant decreases in consumption and treatment expenses. Drug delivery focuses on maximizing bioavailability both at specific places in the body and over a period of time. This can potentially be achieved by molecular targeting by nanoengineered devices. A benefit of using nanoscale for medical technologies is that smaller devices are less invasive and can possibly be implanted inside the body, plus biochemical reaction times are much shorter. These devices are faster and more sensitive than typical drug delivery. The efficacy of drug delivery through nanomedicine is largely based upon: a) efficient encapsulation of the drugs,

b) successful delivery of drug to the targeted region of the body, and c) successful release of the drug.

Drug delivery systems, lipid- or polymer-based nanoparticles, can be designed to improve the pharmacokinetics and biodistribution of the drug. However, the pharmacokinetics and pharmacodynamics of nanomedicine is highly variable among different patients. When designed to avoid the body's defence mechanisms, nanoparticles have beneficial properties that can be used to improve drug delivery. Complex drug delivery mechanisms are being developed, including the ability to get drugs through cell membranes and into cell cytoplasm. Triggered response is one way for drug molecules to be used more efficiently. Drugs are placed in the body and only activate on encountering a particular signal. For example, a drug with poor solubility will be replaced by a drug delivery system where both hydrophilic and hydrophobic environments exist, improving the solubility. Drug delivery systems may also be able to prevent tissue damage through regulated drug release; reduce drug clearance rates; or lower the volume of distribution and reduce the effect on non-target tissue. However, the biodistribution of these nanoparticles is still imperfect due to the complex host's reactions to nano- and microsized materials and the difficulty in targeting specific organs in the body. Nevertheless, a lot of work is still ongoing to optimize and better understand the potential and limitations of nanoparticulate systems. While advancement of research proves that targeting and distribution can be augmented by nanoparticles, the dangers of nanotoxicity become an important next step in further understanding of their medical uses.

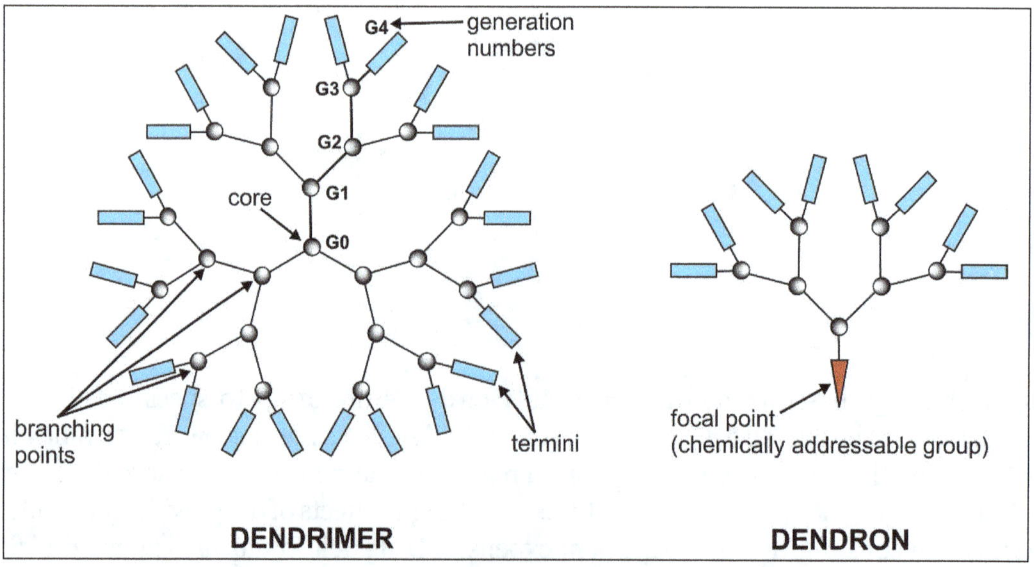

Nanoparticles *(top)*, liposomes *(middle)*, and dendrimers *(bottom)* are some nanomaterials being investigated for use in nanomedicine.

Nanoparticles are under research for their potential to decrease antibiotic resistance or for various antimicrobial uses. Nanoparticles might also be used to circumvent multidrug resistance (MDR) mechanisms.

Systems under Research

Advances in lipid nanotechnology were instrumental in engineering medical nanodevices and novel drug delivery systems, as well as in developing sensing applications. Another system for microRNA delivery under preliminary research is nanoparticles formed by the self-assembly of two different microRNAs deregulated in cancer. One potential application is based on small electromechanical systems, such as nanoelectromechanical systems being investigated for the active release of drugs and sensors for possible cancer treatment with iron nanoparticles or gold shells.

Applications

Some nanotechnology-based drugs that are commercially available or in human clinical trials include:

- Abraxane, approved by the U.S. Food and Drug Administration (FDA) to treat breast cancer, non-small- cell lung cancer (NSCLC) and pancreatic cancer, is the nanoparticle albumin bound paclitaxel.
- Doxil was originally approved by the FDA for the use on HIV-related Kaposi's sarcoma. It is now being used to also treat ovarian cancer and multiple myeloma. The drug is encased in liposomes, which helps to extend the life of the drug that is being distributed. Liposomes are self-assembling, spherical, closed colloidal structures that are composed of lipid bilayers that surround an aqueous space. The liposomes also help to increase the functionality and it helps to decrease the damage that the drug does to the heart muscles specifically.
- Onivyde, liposome encapsulated irinotecan to treat metastatic pancreatic cancer, was approved by FDA in October 2015.
- Rapamune is a nanocrystal-based drug that was approved by the FDA in 2000 to prevent organ rejection after transplantation. The nanocrystal components allow for increased drug solubility and dissolution rate, leading to improved absorption and high bioavailability.

Cancer

Nanoparticles have high surface area to volume ratio. This allows for many functional groups to be attached to a nanoparticle, which can seek out and bind to certain tumor cells. Additionally, the small size of nanoparticles (5 to 100 nanometers), allows them to preferentially accumulate at tumor sites (because tumors lack an effective lymphatic drainage system). Limitations to conventional cancer chemotherapy include drug resistance, lack of selectivity, and lack of solubility.

Imaging

In vivo imaging is another area where tools and devices are being developed. Using nanoparticle contrast agents, images such as ultrasound and MRI have a favorable distribution and improved contrast. In cardiovascular imaging, nanoparticles have potential to aid visualization of blood pooling, ischemia, angiogenesis, atherosclerosis, and focal areas where inflammation is present.

The small size of nanoparticles endows them with properties that can be very useful in oncology, particularly in imaging. Quantum dots (nanoparticles with quantum confinement properties, such as size-tunable light emission), when used in conjunction with MRI (magnetic resonance imaging), can produce exceptional images of tumor sites. Nanoparticles of cadmium selenide (quantum dots) glow when exposed to ultraviolet light. When injected, they seep into cancer tumors. The surgeon can see the glowing tumor, and use it as a guide for more accurate tumor removal.These nanoparticles are much brighter than organic dyes and only need one light source for excitation. This means that the use of fluorescent quantum dots could produce a higher contrast image and at a lower cost than today's organic dyes used as contrast media. The downside, however, is that quantum dots are usually made of quite toxic elements, but this concern may be addressed by use of fluorescent dopants.

Tracking movement can help determine how well drugs are being distributed or how substances are metabolized. It is difficult to track a small group of cells throughout the body, so scientists used to dye the cells. These dyes needed to be excited by light of a certain wavelength in order for them to light up. While different color dyes absorb different frequencies of light, there was a need for as many light sources as cells. A way around this problem is with luminescent tags. These tags are quantum dots attached to proteins that penetrate cell membranes. The dots can be random in size, can be made of bio-inert material, and they demonstrate the nanoscale property that color is size-dependent. As a result, sizes are selected so that the frequency of light used to make a group of quantum dots fluoresce is an even multiple of the frequency required to make another group incandesce. Then both groups can be lit with a single light source. They have also found a way to insert nanoparticles into the affected parts of the body so that those parts of the body will glow showing the tumor growth or shrinkage or also organ trouble.

Sensing

Nanotechnology-on-a-chip is one more dimension of lab-on-a-chip technology. Magnetic nanoparticles, bound to a suitable antibody, are used to label specific molecules, structures or microorganisms. In particular silica nanoparticles are inert from the photophysical point of view and might accumulate a large number of dye(s) within the nanoparticle shell. Gold nanoparticles tagged with short segments of DNA can be used for detection of genetic sequence in a sample. Multicolor optical coding for biological assays has been achieved by embedding different-sized quantum dots into polymeric microbeads. Nanopore technology for analysis of nucleic acids converts strings of nucleotides directly into electronic signatures.

Sensor test chips containing thousands of nanowires, able to detect proteins and other biomarkers left behind by cancer cells, could enable the detection and diagnosis of cancer in the early stages from a few drops of a patient's blood. Nanotechnology is helping to advance the use of arthroscopes, which are pencil-sized devices that are used in surgeries with lights and cameras so surgeons can do the surgeries with smaller incisions. The smaller the incisions the faster the healing time which is better for the patients. It is also helping to find a way to make an arthroscope smaller than a strand of hair.

Research on nanoelectronics-based cancer diagnostics could lead to tests that can be done in pharmacies. The results promise to be highly accurate and the product promises to be inexpensive. They could take a very small amount of blood and detect cancer anywhere in the body in about five minutes, with a sensitivity that is a thousand times better a conventional laboratory test. These devices that are built with nanowires to detect cancer proteins; each nanowire detector is primed to be sensitive to a different cancer marker. The biggest advantage of the nanowire detectors is that they could test for anywhere from ten to one hundred similar medical conditions without adding cost to the testing device. Nanotechnology has also helped to personalize oncology for the detection, diagnosis, and treatment of cancer. It is now able to be tailored to each individual's tumor for better performance. They have found ways that they will be able to target a specific part of the body that is being affected by cancer.

Blood Purification

Magnetic micro particles are proven research instruments for the separation of cells and proteins from complex media. The technology is available under the name Magnetic-activated cell sorting or Dynabeads among others. More recently it was shown in animal models that magnetic nanoparticles can be used for the removal of various noxious compounds including toxins, pathogens, and proteins from whole blood in an extracorporeal circuit similar to dialysis. In contrast to dialysis, which works on the principle of the size related diffusion of solutes and ultrafiltration of fluid across a semi-permeable membrane, the purification with nanoparticles allows specific targeting of substances. Additionally larger compounds which are commonly not dialyzable can be removed.

The purification process is based on functionalized iron oxide or carbon coated metal nanoparticles with ferromagnetic or superparamagnetic properties. Binding agents such as proteins, antibodies, antibiotics, or synthetic ligands are covalently linked to the particle surface. These binding agents are able to interact with target species forming an agglomerate. Applying an external magnetic field gradient allows exerting a force on the nanoparticles. Hence the particles can be separated from the bulk fluid, thereby cleaning it from the contaminants.

The small size (< 100 nm) and large surface area of functionalized nanomagnets leads to advantageous properties compared to hemoperfusion, which is a clinically used technique for the purification of blood and is based on surface adsorption. These advantages are high loading and accessible for binding agents, high selectivity towards the target compound, fast diffusion, small hydrodynamic resistance, and low dosage.

This approach offers new therapeutic possibilities for the treatment of systemic infections such as sepsis by directly removing the pathogen. It can also be used to selectively remove cytokines or endotoxins or for the dialysis of compounds which are not accessible by traditional dialysis methods. However the technology is still in a preclinical phase and first clinical trials are not expected before 2017.

Tissue Engineering

Nanotechnology may be used as part of tissue engineering to help reproduce or repair or reshape damaged tissue using suitable nanomaterial-based scaffolds and growth factors. Tissue engineering if successful may replace conventional treatments like organ transplants or artificial implants. Nanoparticles such as graphene, carbon nanotubes, molybdenum disulfide and tungsten disulfide are being used as reinforcing agents to fabricate mechanically strong biodegradable polymeric nanocomposites for bone tissue engineering applications. The addition of these nanoparticles in the polymer matrix at low concentrations (~0.2 weight %) leads to significant improvements in the compressive and flexural mechanical properties of polymeric nanocomposites. Potentially, these nanocomposites may be used as a novel, mechanically strong, light weight composite as bone implants.

For example, a flesh welder was demonstrated to fuse two pieces of chicken meat into a single piece using a suspension of gold-coated nanoshells activated by an infrared laser. This could be used to weld arteries during surgery. Another example is nanonephrology, the use of nanomedicine on the kidney.

Medical Devices

Neuro-electronic interfacing is a visionary goal dealing with the construction of nanodevices that will permit computers to be joined and linked to the nervous system. This idea requires the building of a molecular structure that will permit control and detection of nerve impulses by an external computer. A refuelable strategy implies energy is refilled continuously or periodically with external sonic, chemical, tethered, magnetic, or biological electrical sources, while a nonrefuelable strategy implies that all power is drawn from internal energy storage which would stop when all energy is drained. A nanoscale enzymatic biofuel cell for self-powered nanodevices have been developed that uses glucose from biofluids including human blood and watermelons. One limitation to this innovation is the fact that electrical interference or leakage or overheating from power consumption is possible. The wiring of the structure is extremely difficult because they must be positioned precisely in the nervous system. The structures that will provide the interface must also be compatible with the body's immune system.

Cell Repair Machines

Molecular nanotechnology is a speculative subfield of nanotechnology regarding the possibility of engineering molecular assemblers, machines which could re-order matter at a molecular or atomic scale. Nanomedicine would make use of these nanorobots, introduced into the body, to repair or detect damages and infections. Molecular nanotechnology is highly theoretical, seeking to anticipate what inventions nanotechnology might yield and to propose an agenda for future inquiry. The proposed elements of molecular nanotechnology, such as molecular assemblers and nanorobots are far beyond current capabilities. Future advances in nanomedicine could give rise to life extension through the repair of many processes thought to be responsible for aging. K. Eric Drexler, one of the founders of nanotechnology, postulated cell repair machines, including ones operating within cells and utilizing as yet hypothetical molecular machines, in his 1986 book *Engines of Creation*, with the first technical discussion of medical nanorobots by Robert Freitas appearing in 1999. Raymond Kurzweil, a futurist and transhumanist, stated in his book *The Singularity Is Near* that he believes that advanced medical nanorobotics could completely remedy the effects of aging by 2030. According to Richard Feynman, it was his former graduate student and collaborator Albert Hibbs who originally suggested to him the idea of a *medical* use for Feynman's theoretical micromachines.

NANOARCHITECTURES FOR LITHIUM-ION BATTERIES

Nanoarchitectures for lithium-ion batteries are attempts to employ nanotechnology to improve the design of lithium-ion batteries. Research in lithium-ion batteries focuses on improving energy density, power density, safety, durability and cost.

Energy Density

Increased energy density requires inserting/extracting more ions from the electrodes. Electrode capacities are compared through three different measures: capacity per unit of mass (known as "specific energy" or "gravimetric capacity"), capacity per unit volume ("volumetric capacity"), and area-normalized specific capacity ("areal capacity").

Power Density

Separate efforts focus on improving power density (rate of charge/discharge). Power density is based upon mass and charge transport, electronic and ionic conductivity, and electron-transfer kinetics; easy transport through shorter distance and greater surface area improve the rates.

Anodes

Carbon anodes are traditionally used because of lithium's ability to intercalate without unacceptable volumetric expansion. The latter damages the battery and reduces the amount of lithium available for charging. Reduced intercalation limits capacity. Carbon based anodes have a gravimetric capacity of 372 mAh/g for LiC_6.

The specific capacity of silicon is approximately ten times greater than carbon. The atomic radius of Si is 1.46 angstroms, while the atomic radius of Li is 2.05 angstroms. The formation of $Li_{3.75}Si$ causes significant volumetric expansion, progressively destroying the anode. Reducing the anode architecture to the nanoscale offers advantages, including improved cycle life and reduced crack propagation and failure. Nanoscale particles are below the critical flaw size within a conductive binder film. Reducing transport lengths(the distance between the anode and cathode) reduces ohmic losses (resistance).

Nanostructuring increases the surface area to volume ratio, which improves both energy and power density due to an increase in the electrochemically active area and a reduction in transport lengths. However, the increase also increases side reactions between the electrode and the electrolyte, causing higher self-discharge, reduced charge/discharge cycles and lower calendar life. Some recent work focused on developing materials that are electrochemically active within the range where electrolyte decomposition or electrolyte/electrode reactions do not occur.

Nonconventional Architectures

A research concept has been proposed, in which the major parts of lithium-ion batteries, that is, anode, electrolyte and cathode are combined in one functional molecule. A layer of such functional molecules aligned by the use of Langmuir-Blodgett method than placed in between two current collectors. The feasibility is not confirmed yet.

Nanostructured Architectures

A significant majority of battery designs are two–dimensional and rely on layered construction. Recent research has taken the electrodes into three-dimensions. This allows for significant improvements in battery capacity; a significant increase in areal capacity occurs between a 2d thick film electrode and a 3d array electrode.

Three-dimensional Thin–films

Solid state batteries employ geometry most similar to traditional thin-film batteries.

Three-dimensional thin-films use the third dimension to increase the electrochemically active area. Thin film two dimensional batteries are restricted to between 2-5 micrometres, limiting areal capacity to significantly less than that of three-dimensional geometries.

Dimensionality is increased by using a perforated substrate. One way to create perforations is through inductive coupled plasma etching on silicon.

Another approached used highly anisotropic etching of a silicon substrate through electrochemical or reactive ion etching to create deep trenches. The requisite layers, an anode, separator, and cathode, for a battery were then added by low-pressure chemical vapor deposition. The battery consists of a thin active silicon layer separated from a thin cathodic layer by a solid-state electrolyte. The electrochemically active area consists of 50 nm nanoparticles, smaller than the critical size for crack propagation.

Interdigitated Electrodes

Another architecture is a periodic grouping of anodic and cathodic poles. For this design power and energy density is maximized by minimizing electrode separation. An innate non-uniform current density occurs and lowers cell efficiencies, reduces stability and produces non-uniform heating within the cell. Relative to a two dimensional battery the length (L) over which transport must occur is decreased by two-thirds, which improves kinetics and reduces ohmic loses. Optimization of L can lead to significant improvement in areal capacity; an L on the size scale of 500 micrometres results in a 350% increase in capacity over a comparable two dimensional battery. However, ohmic losses increase with L, eventually offsetting the enhancement achieved through increasing L.

For this geometry, four main designs were proposed: rows of anodes and cathodes, alternating anodes and cathodes, hexagonally packed 1:2 anodes: Cathodes, and alternating anodic and cathodic triangular poles where the nearest neighbors in the row are rotated 180 degrees.

The row design has a large, non-uniform current distribution. The alternating design exhibits better uniformity, given a high number of electrodes of opposite polarity. For systems with an anode or cathode that is sensitive to non-uniform current density, non-equal numbers of cathodes and anodes can be used; the 2:1 hexagonal design allows for a uniform current density at the anode but a non-uniform current distribution at the cathode. Performance can be increased through changing the shape of the poles. The triangular design improves cell capacity and power by sacrificing current uniformity. A similar system uses interdigitated plates instead of poles.

In 2013 researchers used additive manufacturing to create stacked, interdigitated electrodes. The battery was no larger than a grain of sand. The process placed anodes and cathodes closer to each other than before. The ink for the anode was nanoparticles of one lithium metal oxide compound, and the ink for the cathode from nanoparticles of another. The printer deposited the inks onto the teeth of two gold combs, forming an interlaced stack of anodes and cathodes.

Concentric Electrodes

The concentric cylinder design is similar to interdigitated poles. Instead of discrete anode and cathode poles, the anode or cathode is kept as a pole that is coated by electrolyte. The other electrode serves as the continuous phase in which the anode/cathode resides. The main advantage is that the amount of electrolyte is reduced, increasing energy density. This design maintains a short transport distance like the interdigitated system and thus has a similar benefit to charge and mass transport, while minimizing ohmic loses.

Inverse Opal

A version of the concentric cylinder packed particles or close-packed polymer to create a three-dimensionally ordered macroporous (3DOM) carbon anode. This system is fabricated by using colloidal crystal templating, electrochemical thin-film growth, and soft sol–gel chemistry. 3DOM materials have a unique structure of nanometer thick walls that surround interconnected and closed-packed sub-micrometer voids. The 3DOM structure is coated with a thin polymer layer and then filled with second conducting phase. This method leads to a battery with short transport lengths, high ionic conductivity and reasonable electrical conductivity. It removes the need for additives that do not contribute to electrochemical performance. Performance can be improved by coating with tin oxide nanoparticles to enhance the initial capacity. The coating infiltrates the network formed by the 3DOM structure to produce uniform thickness.

Nanowires and Nanotubes

Nanowire and nanotubes have been integrated with various battery components. The reason for this interest is because of shortened transport lengths, resistance to degradation and storage. For carbon nanotubes (CNT), lithium-ions can be stored on the exterior surface, in the interstitial sites between the nanotubes and on the tube's interior.

Nanowires have been incorporated into the anode/cathode matrix to provide a builtin conductive charge collector and enhancing capacity. The nanowires were incorporated through a solution-based method that allows the active material to be printed on a substrate.

Another approach uses a CNT-cellulose composite. CNTs were grown on a silicon substrate by thermal-CVD and then embedded in cellulose. Finally a lithium electrode is added on top of the cellulose across from the CNTs.

In 2007 Si nanowires were fabricated on a steel substrate by a vapor-liquid solid growth method. These nanowires exhibited close to the theoretical value for silicon and showed only minimal fading after a 20% drop between the first to second cycles. This performance is attributed to the facile strain relaxation that allows for accommodations of large strains, while maintaining good contact with the current collector and efficient 1D electron transport along the nanowire.

Aperoidic Electrodes

Periodic structures lead to non-uniform current densities that lower efficiency and decrease stability. The aperiodic structure is typically made of either aerogels or somewhat more dense ambigels that forms a porous aperiodic sponge. Aerogels and ambigels are formed from wet gels; aerogels are formed when wet gels are dried such that no capillary forces are established, while ambigels are wet gels dried under conditions that minimize capillary forces. Aerogels and ambigels are unique in that 75-99% of the material is 'open' but interpenetrated by a solid that is on the order of 10 nm, resulting in pores on the order of 10 to 100 nm. The solid is covalently networked and resistant to agglomeration and sintering. Beyond aperiodicity, these structures are used because the porous structure allows for rapid diffusion throughout the material, and the porous structure provides a large reaction surface. Fabrication is through coating the ambigel with a polymer electrolyte and then filling the void space with RuO_2 colloids that act as an anode.

Conformal Coatings

Most designs were half-cell experiments; testing only the anode or cathode. As geometries become more complex, non-line-of-sight methods to in-fill the design with electrolyte materials supply the oppositely charged electrode is essential. These batteries can be coated with various materials to improve their performance and stability. However, chemical and physical heterogeneity leaves molecular-level control a significant challenge, especially since the electrochemistry for energy storage is not defect-tolerant.

Layer-by-layer (LbL)

LbL approaches are used to coat 3d nanoarchitecture. Electrostatically binding a charged polymer to an oppositely charged surface coats the surface with polymer. Repeated steps of oppositely charged polymer build up a well-controlled thick layer. Polyelectrolyte films and ultrathin (less than 5 nm) of electroactive polymers have been deposited on planar substrates using this method. However, problems exist with the deposition of polymers within complex geometries, e.g. pores, on the size scale of 50-300 nm, resulting in defective coatings. One potential solution is to use self-limiting approaches.

Atomic Layer Deposition (ALD)

Another approach to coating is ALD which coats the substrate layer-by-layer with atomic precision. The precision is because reactions are confined to the surface containing an active chemical moiety that reacts with a precursor; this limits thickness to one monolayer. This self-limiting growth is essential for complete coatings since deposition does not inhibit the access by other polymeric units to non-coated sites. Thicker samples can be produced by cycling gases in a similar manner to alternating with oppositely charged polymers in LbL. In practice ALD may require a few cycles in order to achieve the desired coverage and can result in varied morphologies such as islands, isolated crystallites, or nanoparticles. Morphology can alter electrochemical behavior and therefore must be carefully controlled.

ALD was also used to deposit iron oxide on 3DOM carbon to enhance reactivity between lithium and oxygen. The iron was then coatedwith palladium nanoparticles, which effectively reduced carbon's destructive reaction with oxygen and improved the discharge cycle. Wang said the findings show 3DOm carbon can meet new performance standards when it is stabilized.

Electropolymerization

Electropolymerization supplies a thin polymer film, 10 to 100 nm. The electropolymerization of an insulating polymer results in self-limiting deposition as the active moiety is protected; the deposition can also be self-limiting if the polymer can block the solubilized monomer and prohibit continued growth. Through the control of electrochemical variables, polyaniline and polythiophene can be deposited in a controlled manner. Styrene, methyl methacrylate, phenols and other electrically insulating polymers have been deposited on the electrodes to act as a separator that allows ionic transport, but inhibits electrical transport to prevent shorts. Mesoporous manganese dioxide ambigels have been protected by 7-9 nm films of polymer such that dissolution of the manganese dioxide in aqueous acid was avoided. Uniform coatings require the architecture to be wetted by the monomer solution; this can be achieved through a solution that displays a similar surface energy to that of the porous solid. As the scale continuous to decrease and transport through the solid becomes more difficult, pre-equilibration is needed to ensure coating uniformity.

NANOTECHNOLOGY IN WARFARE

Nanotechnology in warfare is a branch of nano-science in which molecular systems are designed, produced and created to fit a nano-scale (1-100 nm). The application of such technology, specifically in the area of warfare and defence, has paved the way for future research in the context of weaponisation. Nanotechnology unites a variety of scientific fields including material science, chemistry, physics, biology and engineering.

Advancements in this area, have led to categorised development of such nano-weapons with classifications varying from; small robotic machines, hyper-reactive explosives, and electromagnetic super-materials. With this technological growth, has emerged implications of associated risks and repercussions, as well as regulation to combat these effects. These impacts give rise to issues concerning global security, safety of society, and the environment. Legislation may need to be constantly monitored to keep up with the dynamic growth and development of nano-science, due to the potential benefits or dangers of its use. Anticipation of such impacts through regulation, would 'prevent irreversible damages' of implementing defence related nanotechnology in warfare.

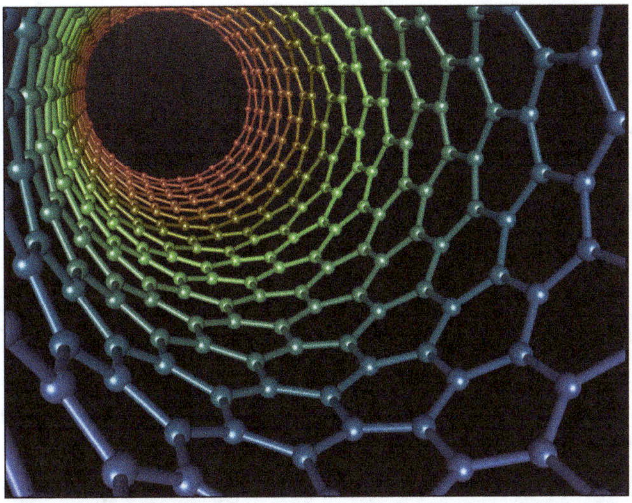

Single-walled Carbon Nanotube, consisting of a graphite sheet wrapped into a cylinder shape.

Historical use of nanotechnology in the area of warfare and defence has been rapid and expansive. Over the past two decades, numerous countries have funded military applications of this technology including; China, United Kingdom, Russia, and most notably the United States. The US government has been considered a national leader of research and development in this area, however now rivalled by international competition as appreciation of nanotechnology's eminence increases. Growth of this sphere, therefore has a dominant platform at the forefront of military interests in the use, or misuse of its power.

In 2000, the United States government developed a National Nanotechnology Initiative to focus funding towards the development of nano-science and its technology, with a heavy focus on utilising the potential of nano-weapons. This initial US proposal has now grown to coordinate application of nanotechnology in numerous defence programs, as well as all military factions including Air Force, Army and Navy. From the financial year 2001 through to 2014, the US government contributed around $19.4 billion to nano-science, moreover the development and manufacturing of nano-weapons for military defence. The 21st Century Nanotechnology Research and Development Act, envisions the United States continuing its leadership in the field of nanotechnology through national collaboration, productivity and competitiveness, to maintain this dominance.

Developments

Successful transitions of nanotechnology into defence products:

- Lifetime of material coatings increased from hours to years, however further development continuing.
- Nano-structured silicate manipulation reducing insulation weight by 980 lbs.
- High Power Microwave (HPM) devices with reduced weight, shape and power consumption.

The United States government has had military purposed development of nanotechnology at the forefront of its national budget and policy throughout the Clinton and Bush administrations, with the Department of Defense planning to continue with this priority throughout the 21st century. In response to America's assertive public funding of defence purposed nanotechnology, numerous global actors have since created similar programmes.

In the sub-category of nano materials China secures second place behind the United States in the amount of research publications they have released. Conjecture stands over the purpose of China's quick development to rival the U.S., with 1/5 of their government budget spent on research (US$337million). In 2018, Tsinghua University, Beijing, released their findings where they have enhanced carbon nanotubes to now withstand the weight of over 800 tonnes, requiring just 1 cm^3 of material. The scientific nanotechnology team hinted at aerospace, and armour boosting applications, showing promise for defence related nano-weapons. The Chinese Academy of Science's Vice President Chunli Bai, has stated the need to focus on closing the gap between "basic research and application," in order for China to advance its global competitiveness in nanotechnology.

Between 2001 and 2004, approximately 60 countries globally implemented national nanotechnology programmes. According to R.D Shelton, an international technology assessor, research and development in this area "has now become a socio-economic target an area of intense international collaboration and competition." As of 2017, data showed 4725 patents published in USPTO by the USA alone, maintaining their position as a leader in nanotechnology for over 20 years.

Most recent research into military nanotechnological weapons includes production of defensive military apparatus, with objectives of enhancing existing designs of lightweight, flexible and durable materials. These innovative designs are equipped with features to also enhance offensive strategy through sensing devices and manipulation of electromechanical properties.

Soldier Battlesuit

The Institute for Soldier Nanotechnologies (ISN), deriving from a partnership between the United States Army and MIT, provided an opportunity to focus funding and research activities purely on developing armour to increase soldier survival. Each of seven teams produces innovative enhancements for different aspects of a future U.S. soldier body-suit. These additional characteristics include energy-absorbing material protecting from blasts or ammunition shocks, engineered sensors to detect chemicals and toxins, as well as built in nano devices to identify personal medical issues such as haemorrhages and fractures. This suit would be made possible with advanced nano-materials such as carbon nanotubes woven into fibres, allowing strengthened structural capacities and flexibility, however preparation becomes an issue due to inability to use automated manufacturing.

Enhanced Materials

Creation of sol-gel ceramic coatings has protected metals from; wear, fractures and moisture, allowing adjustability to numerous shapes and sizes, as well as aiding "materials that cannot withstand high temperature". Current research focuses on resolving durability issues, where stress cracks between the coating and material set limitations on its use and longevity. The drive for this research is finding more efficient and cost effective uses in application of nanotechnology for Airforce and Navy military groups. Integration of fibre-reinforced nano-materials in structural features, such as missile casings, can limit overheating, increase reliability, strength and ductility of the materials used for such nanotechnology.

Communication Devices

Nanotechnology designed for advanced communication is expected to equip soldiers and vehicles with micro antenna rays, tags for remote identification, acoustic arrays, micro GPS receivers and wireless communication. Nanotech facilitates easier defence related communications due to lower energy consumption, light weight, efficiency of power, as well as smaller and cheaper to manufacture. Specific military uses of this technology include aerospace applications such as; solid oxide fuel cells to provide three times the energy, surveillance cameras on microchips, performance monitors, and cameras as light as 18g.

Mini-nukes

The United States, along with countries such as Russia and Germany, are utilising the convenience of small nanotechnologies, adhering it to nuclear "mini-nuke" explosive devices. This weapon would weigh 5 lbs, with the force of 100 tonnes of TNT, giving it the possibility to annihilate and threaten humanity. The structural integrity would remain the same as nuclear bombs, however manufactured with nano-materials to allow production to a smaller scale.

Engineers and scientists alike, realise some of these proposed developments may not be feasible within the next two decades as more research needs to be undertaken, improving models to be quicker and more efficient. Particularly molecular nanotechnology, requires further understanding of manipulation and reaction, in order to adapt it to a military arena.

Implications

Nanotechnology and its use in warfare promises economic growth however comes with the increased threat to international security and peacekeeping. The rapid emergence of new nanotechnologies have sparked discussion surrounding the impacts such developments will have on geo-politics, ethics, and the environment.

Geo-political

Difficulty in categorisation of nano-weapons, and their intended purposes (defensive or offensive) compromises the balance of stability and trust in the global environment. "A lack of transparency about an emerging technology not only negatively effects public perception but also negatively impacts the perceived balance of powers in the existing security environment." The peace and cohesion of the international structure may possibly be negatively affected with a continuing military-focused development of nanotechnology in warfare. Ambiguity and a lack of transparency in research increases difficulty of regulation in this area. Similarly, arguments put forward from a scientific standpoint, highlight the limited information known, concerning the implications of creating such powerful technology, in regards to reaction of the nano-particles themselves. "Although great scientific and technological progress has been made, many questions about the behaviour of matter at the nanoscale level remain, and considerable scientific knowledge has yet to be learned."

Environmental

The introduction of nanotechnology into everyday life enables potential benefits of use, yet carries the possibility of unknown consequences for the environment and safety. Possible positive developments include creation of nano-devices to decrease remaining radio-activity in areas, as well as sensors to detect pollutants and adjust fuel-air mixtures. Associated risks may involve; military personnel inhaling nanoparticles added to fuel, possible absorption of nanoparticles from sensors into the skin, water, air or soil, dispersion of particles from blasts through the environment (via wind), alongside disposal of nano-tech batteries potentially affecting ecosystems. Applications for materials or explosive devices, allow a greater volume of nano-powders to be packed into a smaller weapon, resulting in a stronger and possibly lethal toxic effect.

Social and Ethical

It is unknown the full extent of consequences that may arise in social and ethical areas. Estimates can be made on the associated impacts as they may mirror similar progression of technological developments and affect all areas. The main ethical uncertainties entail the degree to which modern nanotechnology will threaten privacy, global equity and fairness, while giving rise to patent and property right disputes. An overarching social and humanitarian issue, branches from the creative intention of these developments. 'The power to kill or capture debate', highlights the unethical purpose and function of destruction these nanotechnological weapons supply to the user.

Controversy surrounding the innovation and application of nanotechnology in warfare highlights dangers of not pre-determining risks, or accounting for possible impacts of such technology. "The threat of nuclear weapons led to the cold war. The same trend is foreseen with nanotechnology, which may lead to the so-called nanowars, a new age of destruction", stated by the U.S. Department of Defense. Similarly a report released by Oxford University, warns of the pre-eminent extinction of the human race with a 5% risk of this occurring due to development of 'molecular nanotech weapons'.

APPLICATION OF NANOTECHNOLOGY IN SPACE

Nanotechnology will play an important role in future space missions. Nanosensors, dramatically improved high-performance materials, or highly efficient propulsion systems are but a few examples.

Propulsion Systems

Most of today's rocket engines rely on chemical propulsion. All current spacecraft use some form of chemical rocket for launch and most use them for attitude control as well (the control of the angular position and rotation of the spacecraft, either relative to the object that it is orbiting, or relative to the celestial sphere). Real rocket scientists though are actively researching new forms of space propulsion systems.

One heavily researched area is electric propulsion (EP) that includes field emission electric propulsion (FEEP), colloid thrusters and other versions of field emission thrusters (FETs). EP systems significantly reduce the required propellant mass compared to conventional chemical rockets, allowing to increase the payload capacity or decrease the launch mass. EP has been successfully demonstrated as primary propulsion systems for NASA's Deep Space 1, Japan's HAYABUSA, and ESA's SMART-1 missions.

A nanotechnology EP concept proposes to utilize electrostatically charged and accelerated nanoparticles as propellant. Millions of micron-sized nanoparticle thrusters would fit on one square centimeter, allowing the fabrication of highly scalable thruster arrays.

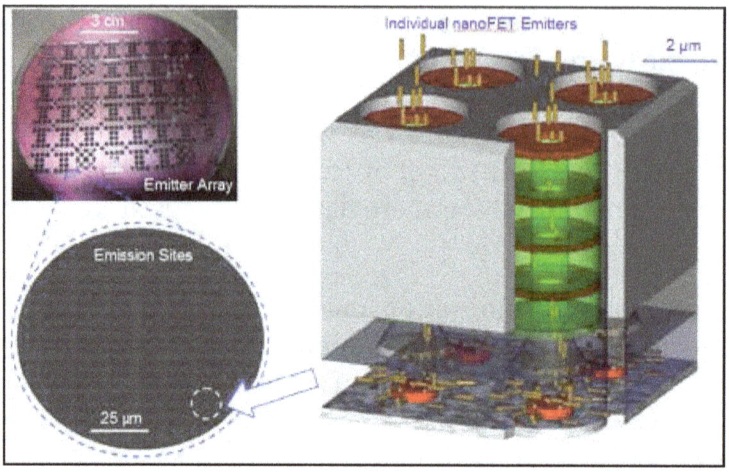

NanoFET characteristic size scales.

Pretty far out are proposals that the manipulation of Casimir forces could lead to a propulsion system for interstellar spaceships. The basic idea is that if one could exploit the fact that vacuum is an energy reservoir, thanks to zero-point energy, future space travelers would have access to a limitless energy source. The only thing they need, of course, is some kind of propulsion system that harvests the required energy from the vacuum. That this is not totally crazy was demonstrated in a 1984 paper. Serious research efforts are being made in various laboratories to harness the Casimir and related effects for vacuum energy conversion.

Radiation Shielding

Radiation shielding is an area where nanotechnology could make a major contribution to human space flight. NASA says that the risks of exposure to space radiation are the most significant factor limiting humans' ability to participate in long-duration space missions. A lot of research therefore focuses on developing countermeasures to protect astronauts from those risks. To meet the needs for radiation protection as well as other requirements such as low weight and structural stability, spacecraft designers are looking for materials that help them develop multifunctional spacecraft hulls.

Advanced nanomaterials such as the newly developed, isotopically enriched boron nanotubes could pave the path to future spacecraft with nanosensor-integrated hulls that provide effective radiation shielding as well as energy storage.

Another area of required radiation shielding is the protection of onboard electronics. It has been reported previously that electronic devices became more radiation tolerant when their dimensions are reduced. For example, multi-quantum well or quantum dot devices can be tens or hundreds times more radiation tolerant than conventional bulk devices. It even was shown that quantum dot/CNT-based photovoltaic devices were five orders of magnitude more resistant than conventional bulk solar cells.

Recently, a few studies on radiation effects of high energetic particles such as proton, electron, and heavy ions on nanomaterials like carbon nanotubes and nanowires have focused on the changed structural properties of the nanomaterials after irradiation.

Anti-satellite Weapon Countermeasure

In January 2007, China successfully tested an Anti-satellite (ASAT) missile system by destroying their own defunct LEO satellite, which generated huge amounts of space debris. This ASAT test raised worldwide concerns about the vulnerability of satellites and other space assets and possibility of triggering an arms race in space.

In order to meet emerging challenges posed by such ASAT missile systems, military strategists and researchers are developing novel technologies to protect their space assets. In view of this, Raytheon Company has developed a counter measure system using quantum dots to protect space assets such as satellites from missile attacks. They have developed a decoy consisting of quantum dots of different sizes and shapes that are engineered to emit radiation having a radiation profile similar to that of the asset.

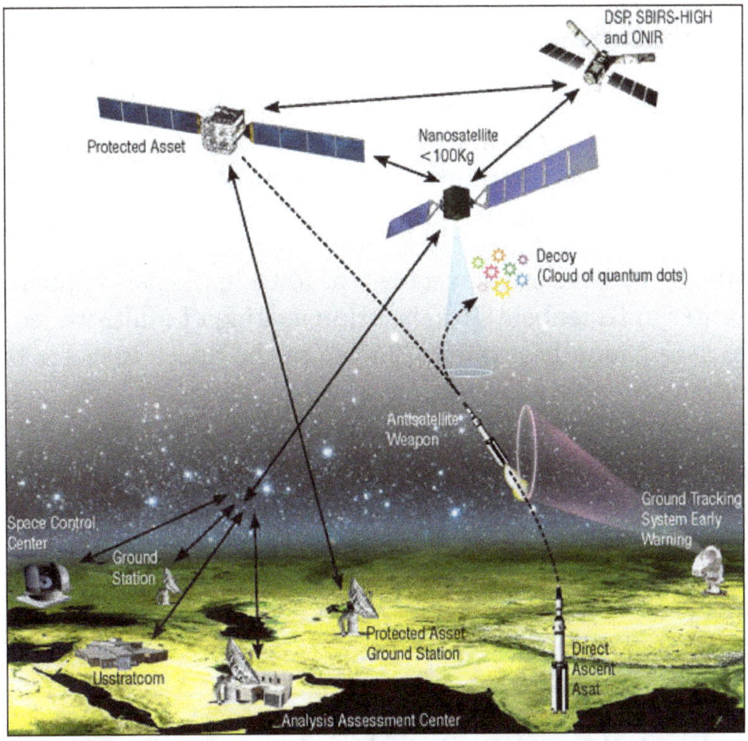

Schematic of a countermeasure system: Anti-satellite weapon seeks the decoy cloud of quantum dots instead of the target satellite.

Space Elevator

Tie a rock to the end of a piece of ribbon, then spin it over your head. It will be pulled taut as the rock circles about. Now, imagine a ribbon 62,000 miles long, anchored near the equator with a weight on the other end. The centrifugal force of the earth's rotation will make it behave the same way. You'll end up with not only the world's biggest nunchuck, but also a kind of elevator to outer space.

A space elevator based on carbon nanotube cables is one of those ideas from 1950s-style futurism that are so whacky they might just work.

There is an annual Space Elevator Conference to discuss progress and a Japanese construction firm even has announced plans to build a space elevator by 2050.

Protecting Satellites from Energy Weapons

A recent research discusses how nanotechnology can be used to improve the design of satellites to mitigate the threats posed by ground-based directed energy weapons and high-powered microwaves. The paper states that several nations, including the U.S., Russia and China, already have either built or are developing the technology to construct ground-based directed energy weapons.

Space Instrumentation

Black is black, right? Not so, according to a team of NASA engineers now developing a blacker-than pitch nanomaterial that will help scientists gather hard-to-obtain scientific measurements or observe currently unseen astronomical objects, like Earth-sized planets in orbit around other stars.

The nanomaterial being developed by a team of 10 technologists at the NASA Goddard Space Flight Center in Greenbelt, Md., is a thin coating of multi-walled carbon nanotubes. While carbon nanotubes would find use in the Space Elevator thanks to their extraordinary strength, in this application, NASA is interested in using the technology to help suppress errant light that has a funny way of ricocheting off instrument components and contaminating measurements.

ENERGY APPLICATIONS OF NANOTECHNOLOGY

An important subfield of nanotechnology related to energy is nanofabrication. Nanofabrication is the process of designing and creating devices on the nanoscale. Creating devices smaller than 100 nanometers opens many doors for the development of new ways to capture, store, and transfer energy. The inherent level of control that nanofabrication could give scientists and engineers would be critical in providing the capability of solving many of the problems that the world is facing today related to the current generation of energy technologies.

People in the fields of science and engineering have already begun developing ways of utilizing nanotechnology for the development of consumer products. Benefits already observed from the design of these products are an increased efficiency of lighting and heating, increased electrical storage capacity, and a decrease in the amount of pollution from the use of energy. Benefits such as these make the investment of capital in the research and development of nanotechnology a top priority.

Consumer Products

Recently, previously established and entirely new companies such as BetaBatt, Inc. and Oxane Materials are focusing on nanomaterials as a way to develop and improve upon older methods for the capture, transfer, and storage of energy for the development of consumer products.

ConsERV, a product developed by the Dais Analytic Corporation, uses nanoscale polymer membranes to increase the efficiency of heating and cooling systems and has already proven to be a lucrative design. The polymer membrane was specifically configured for this application by selectively engineering the size of the pores in the membrane to prevent air from passing, while allowing moisture to pass through the membrane. ConsERV's value is demonstrated in the form of an energy recovery a device which pretreats the incoming fresh air to a building using the energy found in the exhaust air steam using no moving parts to lower the energy and carbon footprint of existing forms of heating and cooling equipment Polymer membranes can be designed to selectively allow particles of one size and shape to pass through while preventing other. This makes for a powerful tool that can be used in all markets - consumer, commercial, industrial, and government products from biological weapons protection to industrial chemical separations. Dais's near term uses of this 'family' of selectively engineered nanotechnology materials, aside from ConsERV, include (a.) a completely new cooling cycle capable of replacing the refrigerant based cooling cycle the world has known for the past 100 plus years. This product, under development, is named NanoAir. NanoAir uses only water and this selectively engineered membrane material to cool (or heat) and dehumidify (or humidify) air. There are no fluorocarbon producing gasses used, and the energy required to cool a space drops as thermodynamics does the actual cooling. The company was awarded an Advanced Research Program Administration - Energy award in 2010, and a United States Department of Defense (DoD) grant in 2011 both designed to accelerate this newer, energy efficient technology closer to commercialization, and (b.) a novel way to clean most all contaminated forms of water called NanoClear. By using the selectivity of this hermetic, engineered composite material it can transfer only a water molecule from one face of the membrane to the other leaving behind the contaminants. It should also be noted Dais received a US Patent in October 2011 titled "Nanoparticle Ultracapacitor". This patented item again uses the selectively engineered material to create an energy storage mechanism projected to have performance and cost advantages over existing storage technologies. The company has used this patent's concepts to create a functional energy storage prototype device named NanoCap. NanoCap is a form of ultra-capacitor potentially useful to power a broad range of applications including most forms of transportation, energy storage (especially useful as a storage media for renewable energy technologies), telecommunication infrastructure, transistor gate dielectrics, and consumer battery applications (cell phones, computers, etc.).

A New York-based company called Applied NanoWorks, Inc. has been developing a consumer product that utilizes LED technology to generate light. Light-emitting diodes or LEDs, use only about 10% of the energy that a typical incandescent or fluorescent light bulb uses and typically last much longer, which makes them a viable alternative to traditional light bulbs. While LEDs have been around for decades, this company and others like it have been developing a special variant of LED called the white LED. White LEDs consist of semi-conducting organic layers that are only about 100 nanometers in distance from each other and are placed between two electrodes, which create an anode, and a cathode. When voltage is applied to the system, light is generated when electricity passes through the two organic layers. This is called electroluminescence. The semiconductor properties of the organic layers are what allow for the minimal amount of energy necessary to generate light. In traditional light bulbs, a metal filament is used to generate light when electricity is run through the filament. Using metal generates a great deal of heat and therefore lowers efficiency.

Reduction of Energy Consumption

A reduction of energy consumption can be reached by better insulation systems, by the use of more efficient lighting or combustion systems, and by use of lighter and stronger materials in the transportation sector. Currently used light bulbs only convert approximately 5% of the electrical energy into light. Nanotechnological approaches like or quantum caged atoms (QCAs) could lead to a strong reduction of energy consumption for illumination.

A reduction of energy consumption can be reached by introducing ZnO nano particles to diesel and bio fuels. At a concentration of 8 parts per million ZnO particles provide an additional catalytic surface, creating additional oxygen during engine combustion, converting more fuel to energy, thus improving fuel efficiency. An unavoidable consequence of improving fuel efficiency in this way is a reduction of harmful emissions.

The use of ZnO nano particles particularly in bio fuel reduces injector wetting (sometimes referred to as injector fouling).

Increasing the Efficiency of Energy Production

Today's best solar cells have layers of several different semiconductors stacked together to absorb light at different energies but they still only manage to use 40 percent of the Sun's energy. Commercially available solar cells have much lower efficiencies

(15-20%). Nanostructuring has been used to improve the efficiencies of established photovoltaic technologies, for example by improving current collection in amorphous silicon devices, plasmonic enhancement in dye-sensitized solar cells, and improved light trapping in crystalline silicon. Furthermore, nanotechnology could help increase the efficiency of light conversion by using nanostructures with a continuum of bandgaps, or by controlling the directivity and photon escape probability of photovoltaic devices.

The degree of efficiency of the internal combustion engine is about 30-40% at present. Nanotechnology could improve combustion by designing specific catalysts with maximized surface area. In 2005, scientists at the University of Toronto developed a spray-on nanoparticle substance that, when applied to a surface, instantly transforms it into a solar collector.

Nuclear Accident Clean-up and Waste Storage

Nanomaterials deployed by swarm robotics may be helpful for decontaminating a site of a nuclear accident which poses hazards to humans because of high levels of radiation and radioactive particles. Hot nuclear compounds such as corium or melting fuel rods may be contained in "bubbles" made from nanomaterials that are designed to isolate the harmful effects of nuclear activity occurring inside of them from the outside environment that organisms inhabit.

Economic Benefits

The relatively recent shift toward using nanotechnology with respect to the capture, transfer, and storage of energy has and will continue to have many positive economic impacts on society. The control of materials that nanotechnology offers to scientists and engineers of consumer products is one of the most important aspects of nanotechnology. This allows for an improved efficiency of products across the board.

A major issue with current energy generation is the loss of efficiency from the generation of heat as a by-product of the process. A common example of this is the heat generated by the internal combustion engine. The internal combustion engine loses about 64% of the energy from gasoline as heat and an improvement of this alone could have a significant economic impact. However, improving the internal combustion engine in this respect has proven to be extremely difficult without sacrificing performance. Improving the efficiency of fuel cells through the use of nanotechnology appears to be more plausible by using molecularly tailored catalysts, polymer membranes, and improved fuel storage.

In order for a fuel cell to operate, particularly of the hydrogen variant, a noble-metal catalyst (usually platinum, which is very expensive) is needed to separate the electrons from the protons of the hydrogen atoms. However, catalysts of this type are extremely sensitive to carbon monoxide reactions. In order to combat this, alcohols or hydrocarbons compounds are used to lower the carbon monoxide concentration in the system. This adds an additional cost to the device. Using nanotechnology, catalysts can be designed through nanofabrication that are much more resistant to carbon monoxide reactions, which improves the efficiency of the process and may be designed with cheaper materials to additionally lower costs.

Fuel cells that are currently designed for transportation need rapid start-up periods for the practicality of consumer use. This process puts a lot of strain on the traditional polymer electrolyte membranes, which decreases the life of the membrane requiring frequent replacement. Using nanotechnology, engineers have the ability to create a much more durable polymer membrane, which addresses this problem. Nanoscale polymer membranes are also much more efficient in ionic conductivity. This improves the efficiency of the system and decreases the time between replacements, which lowers costs.

Another problem with contemporary fuel cells is the storage of the fuel. In the case of hydrogen fuel cells, storing the hydrogen in gaseous rather than liquid form improves the efficiency by 5%. However, the materials that we currently have available to us significantly limit fuel storage due to low stress tolerance and costs. Scientists have come up with an answer to this by using a nanoporous styrene material (which is a relatively inexpensive material) that when super-cooled to around -196 °C, naturally holds on to hydrogen atoms and when heated again releases the hydrogen for use.

Capacitors: Then and Now

For decades, scientists and engineers have been attempting to make computers smaller and more efficient. A crucial component of computers are capacitors. A capacitor is a device that is made of a pair of electrodes separated by an insulator that each stores an opposite charge. A capacitor stores a charge when it is removed from the circuit that it is connected to; the charge is released when it is replaced back into the circuit. Capacitors have an advantage over batteries in that they release their charge much more quickly than a battery.

Traditional or foil capacitors are composed of thin metal conducting plates separated by an electrical insulator, which are then stacked or rolled and placed in a casing. The problem with a traditional capacitor such as this is that they limit how small an engineer can design a computer. Scientists and engineers have since turned to nanotechnology for a solution to the problem.

Using nanotechnology, researchers developed what they call “ultracapacitors.” An ultracapacitor is a capacitor that contains nanocomponents. Ultracapacitors are being researched heavily because of their high density interior, compact size, reliability, and high capacitance. This decrease in size makes it increasingly possible to develop much smaller circuits and computers. Ultracapacitors also have the capability to supplement batteries in hybrid vehicles by providing a large amount of energy during peak acceleration and allowing the battery to supply energy over longer periods of time, such as during a constant driving speed. This could decrease the size and weight of the large batteries needed in hybrid vehicles as well as take additional stress off the battery. However, the combination of ultracapacitors and a battery is not cost effective due to the need of additional DC/DC electronics to coordinate the two.

Nanoporous carbon aerogel is one type of material that is being utilized for the design of ultracapacitors. These aerogels have a very large interior surface area and can have its properties altered by changing the pore diameter and distribution along with adding nanosized alkali metals to alter its conductivity.

Carbon nanotubes are another possible material for use in an ultracapacitor. Carbon nanotubes are created by vaporizing carbon and allowing it to condense on a surface. When the carbon condenses, it forms a nanosized tube composed of carbon atoms. This tube has a high surface area, which increases the amount of charge that can be stored. The low reliability and high cost of using carbon nanotubes for ultracapacitors is currently an issue of research.

In study concerning ultracapacitors or supercapacitors, researchers at the Sungkyunkwan University in the Republic of Korea explored the possibility of increasing the capacitance of electrodes through the addition of fluorine atoms to the walls of carbon nanotubes. As briefly mentioned before, carbon nanotubes are an increasing form of capacitors due to their superb chemical stability, high conductivity, light mass, and their large surface area. These researchers fluorinated single-walled carbon nanotubes (SWCNTs) at high temperatures to bind fluorine atoms to the walls. The attached fluorine atoms changed the non-polar nanotubes to become polar molecules. This can be attributed to the charge transfer from the fluorine. This created dipole-dipole layers along the carbon nanotube walls. Testing of these fluorinated SWCNTs against normal state SWCNTs showed a difference in capacitance. It was determined that the fluorinated SWCNTs are advantageous in fabricating electrodes for capacitors and improve the wettability with aqueous electrolytes, which promotes the overall performance of supercapacitors. While this study brought to knowledge a more efficient example of capacitors, little is known about this new supercapacitor, large scale synthesis is lacking and is necessary for any massive production, and preparation conditions are quite tedious in achieving the final product.

Theory of Capacitance

Understanding the concept of capacitance can be helpful in understanding why nanotechnology is such a powerful tool for the design of higher energy storing capacitors. A capacitor's capacitance (C) or amount of energy stored is equal to the amount of charge (Q) stored on each plate divided by the voltage (V) between the plates. Another representation of capacitance is that capacitance (C) is approximately equal to the permittivity (ε) of the dielectric times the area (A) of the plates divided by the distance

(d) between them. Therefore, capacitance is proportional to the surface area of the conducting plate and inversely proportional to the distance between the plates.

Using carbon nanotubes as an example, a property of carbon nanotubes is that they have a very high surface area to store a charge. Using the above proportionality that capacitance (C) is proportional to the surface area (A) of the conducting plate; it becomes obvious that using nanoscaled materials with high surface area would be great for increasing capacitance. The other proportionality described above is that capacitance (C) is inversely proportional to the distance (d) between the plates. Using nanoscaled plates such as carbon nanotubes with nanofabrication techniques, gives the capability of decreasing the space between plates which again increases capacitance.

NANOTECHNOLOGY IN FUEL CELLS

A fuel cell is a device which converts a fuel directly into electricity in an electrochemical reaction. This is in contrast to most methods of generating electricity, which use the heat from burning fuel to generate electricity mechanically.

Fuel cells have the potential to be an incredibly efficient power source. They can theoretically operate on a wide range of fuels, and the technology can be scaled from portable fuel cells in laptops, up to huge stationary installations to power data centres.

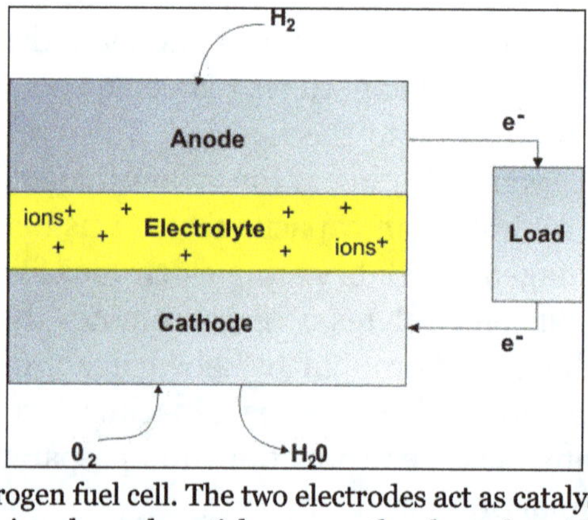

Schematic of a hydrogen fuel cell. The two electrodes act as catalysts for the chemical reaction, as well as passing charged particles across the electrolyte. The electrolyte is usually a conducting solid ceramic, aqueous solution, or molten phosphoric acid.

There are many limitations preventing fuel cells from reaching widespread commercial use, however. Expensive materials such as platinum are needed for the electrode catalysts. Fuels other than hydrogen can cause fouling of the electrodes, and hydrogen is costly to produce and difficult to store. The most efficient types of fuel cell operate at very high temperatures, which reduces their lifespan due to corrosion of the fuel cell components.

Nanotechnology may be able to ease many of these problems. Recent nanotechnology research has produced a number of promising nanomaterials which could make fuel cells cheaper, lighter and more efficient.

Nanotechnology for Fuel Cell Catalysts

The catalytic electrodes in fuel cells are most often made from platinum. It has been known for some time that using platinum nanoparticles instead of a solid platinum surface increases efficiency, and allows much less metal to be used.

One potential improvement to the current technology is to support the platinum nanoparticles on a porous sufrace, such as an activated carbon, or nanostructures like carbon nanotubes or nanowalls. This further increases the accessibility of the platinum surfaces, decreasing the amount of the expensive metal which is needed to make an effective catalytic electrode.

Carbon Nanotube Catalysts

Modified carbon nanotubes may also be able to replace platinum in fuel cells altogether. Fabrication technology for carbon nanotubes is advancing rapidly, and the cheap and abundant raw material means that catalysts based on carbon nanotubes can be produced for a fraction of the cost of platinum catalysts. As platinum currently accounts for at least 25% of the cost of commercial fuel cells, adoption of these catalysts will remove a major barrier to many applications of fuel cells.

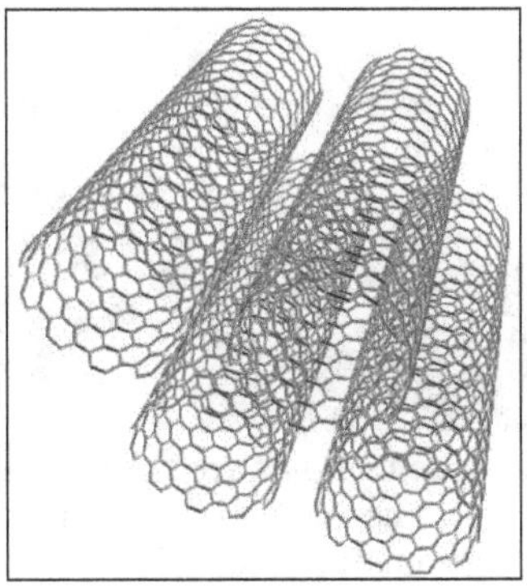

Catalytic electrodes based on modified carbon nanotubes could make fuel cells significantly cheaper and able to cope with more diverse fuels.

By doping carbon nanotubes with nitrogen, or coating them in an electron-withdrawing polymer (polydiallyldimethylammonium chloride, or PDDA), the electronic properties of the nanotubes can be altered so as to make them effective as a catalyst.

The electrocatalytic activity of these modified nanotubes can actually be superior to that of platinum - the power output of a fuel cell using carbon nanotube electrodes is equal to or greater than that from the platinum equivalent.

The nanotube electrodes are also more robust. Their catalytic activity is not damaged by carbon monoxide or the crossover effect when using methanol as the fuel, unlike platinum, which improves the lifetime of the cell.

Self-cleaning Catalysts

Current commercial fuel cells can only run on a limited range of fuels. Most use hydrogen, and some are able to use methanol or natural gas. Whilst other hydrocarbon fuels would be cheaper, they produce carbon monoxide and carbon soot deposits, which poison the electrodes of the fuel cell after a short amount of time. Running at higher temperatures (>900C) can reduce the amount of these poisons which is produced, but that increases the stress on the structure of the cell, requiring costly materials for the interconnects and other important components.

In 2011, researchers at Georgia Tech announced that they could modify the surface of a standard electrode to remove carbon deposits as they form using barium oxide nanoparticles. The nanoparticle coating forms "islands" on the surface, leaving plenty of space for the normal reaction to take place, whilst initiating an oxidation reaction that clears the catalyst surface of any deposits.

Large fuel cell installation. Solid oxide fuel cells are already beginning to be used to efficiently produce energy for large buildings, warehouses and data centers. Waste heat from the fuel cells is captured and used to heat the buildings, resulting in up to 90% efficiency overall.

This development will allow fuel cells to be run on coal gas, produced by gasification of coal, at a much lower temperature than previously, allowing savings on the materials used in the rest of the cell. This is also an incredibly clean way to use coal to generate energy - the exhaust stream is almost pure carbon dioxide, allowing easier operation of carbon sequestration technologies, without any intermediate purification steps.

A similar fuel gas can also be produced by gasification of any carbon-rich material, including biomass such as wood or waste straw. Running fuel cells on these fuels could be a great route to large-scale production of carbon-neutral energy.

Thin-film Electrolyte Membranes

Most commercial fuel cell designs use a solid electrolyte, such as a conducting ceramic like yttria-stabilized zirconia (YSZ), or a polymer like perfluorosulfonic acid. Fuel cells of this type have to operate at very high temperatures. This causes a problem, as the performance of the cell is improved by making the electrolyte layer thinner, reducing its electrical resistance. Very thin electrolyte films, however, suffer mechanical damage at the temperatures required for efficient cell operation.

In 2010, Harvard researchers developed a method of stabilizing a thin film of YSZ, just 100 nm thick. This was accomplished by depositing a micrometer-scale metal grid onto the YSZ film. This method allowed films with lateral dimensions up to the centimetre range. These ultra-thin electrolyte layers could significantly increase the layer density of fuel cells, making the overall size and weight of large installations much more favourable.

References

- Madou, Marc (13 December 2005). "Nanotechnology:dry versus wet engineering?". Analytical and Bioanalytical Chemistry. 4. 384 (6): 4–6. doi:10.1007/s00216-005-0182-7. PMID 16344928
- DNA polyhedra: Goodman, Russel P.; Schaap, Iwan A. T.; Tardin, C. F.; Erben, Christof M.; Berry, Richard M.; Schmidt, C.F.; Turberfield, Andrew J. (9 December 2005). "Rapid chiral assembly of rigid DNA building blocks for molecular nanofabrication". Science. 310 (5754): 1661–1665. Bibcode:2005Sci...310.1661G. doi:10.1126/science.1120367. PMID 16339440
- Nanotechnology-in-space: nanowerk.com, Retrieved 08 February, 2020
- Clarke J; Wu HC; Jayasinghe L; Patel A; Reid A; Bayley H (2009). "Continuous base identification for single-molecule nanopore DNA sequencing". Nature Nanotechnology. 4 (4): 265–270. Bibcode:2009NatNa...4..265C. doi:10.1038/nnano.2009.12. PMID 19350039
- Sierra, D. P.; Weir, N. A.; Jones, J. F. (2005). "A review of research in the field of nanorobotics" (PDF). U.S. Department of Energy – Office of Scientific and Technical Information Oak Ridge, TN. SAND2005-6808: 1–50. doi:10.2172/875622
- Freitas RA (March 2005). "What is nanomedicine?" (PDF). Nanomedicine. 1 (1): 2–9. doi:10.1016/j.nano.2004.11.003. PMID 17292052

6

Nanotechnology and the Environment

Environmental nanotechnology is the branch of nanotechnology that deals with the mitigation and prevention of damage to the environment. Some of its other divisions that aim at improving the environment are green nanotechnology, etc. This chapter discusses nanotechnology in relation with the environment in detail.

ENVIRONMENTAL NANOTECHNOLOGY

Nanotechnology is being used in several applications to improve the environment. This includes cleaning up existing pollution, improving manufacturing methods to reduce the generation of new pollution, and making alternative energy sources more cost effective.

Application of Nanotechnology to Environmental Issues

In trying to help our ailing environment, nanotechnology researchers and developers are pursuing the following avenues:

- Generating less pollution during the manufacture of materials: One example of this is how researchers have demonstrated that the use of silver nanoclusters as catalysts can significantly reduce the polluting byproducts generated in the process used to manufacture propylene oxide. Propylene oxide is used to produce common materials such as plastics, paint, detergents and brake fluid.
- Producing solar cells that generate electricity at a competitive cost: Researcher have demonstrated that an array of silicon nanowires embedded in a polymer results in low cost but high efficiency solar cells. This, or other efforts using nanotechnology to improve solar cells, may result in solar cells that generate electricity as cost effectively as coal or oil.
- Increasing the electricity generated by windmills: Epoxy containing carbon nanotubes is being used to make windmill blades. The resulting blades are

stronger and lower weight and therefore the amount of electricity generated by each windmill is greater.

- Cleaning up organic chemicals polluting groundwater: Researchers have shown that iron nanoparticles can be effective in cleaning up organic solvents that are polluting groundwater. The iron nanoparticles disperse throughout the body of water and decompose the organic solvent in place. This method can be more effective and cost significantly less than treatment methods that require the water to be pumped out of the ground.
- Cleaning up oil spills: Using photocatalytic copper tungsten oxide nanoparticles to break down oil into biodegradable compounds. The nanoparticles are in a grid that provides high surface area for the reaction, is activated by sunlight and can work in water, making them useful for cleaning up oil spills.
- Clearing volatile organic compounds (VOCs) from air: Researchers have demonstrated a catalyst that breaks down VOCs at room temperature. The catalyst is composed of porous manganese oxide in which gold nanoparticles have been embedded.
- Reducing the cost of fuel cells: Changing the spacing of platinum atoms used in a fuel cell increases the catalytic ability of the platinum. This allows the fuel cell to function with about 80% less platinum, significantly reducing the cost of the fuel cell.
- Storing hydrogen for fuel cell powered cars: Using graphene layers to increase the binding energy of hydrogen to the graphene surface in a fuel tank results in a higher amount of hydrogen storage and a lighter weight fuel tank. This could help in the development of practical hydrogen-fueled cars.

GREEN NANOTECHNOLOGY

Nanotechnology is a subject which has been popular within the scientific and technology industries for many years. It is now with the ever growing advancement in technology that nanotechnology is picking up the pace and has got a lot of people talking.

Now, engineers are studying ways that it can be made beneficial to the environment. This has been branded as 'green nanotechnology' since it focuses on challenges within the nanoscale that need to be overcome to ensure eco-friendly processes and products. The objectives of nanotechnology are to create eco-friendly designs with nanotechnology and use it to reduce health and environmental hazards by seeking methods to replace present applications with green nanotechnology products.

Green nanotechnology involves the following:

- Use of less energy during manufacture.
- Ability to recycle after use.
- Using eco-friendly materials.

Nanotechnologies will not just initiate the next industrial revolution, it will also offer technological solutions. It is expected that green nanotechnologies will assist in the following:

- It will help provide clean water to billions through novel filtration techniques and the capability to decontaminate dirty water.
- A large number of efficiency issues stopping the extensive use of renewable energy generation can be solved using nanotechnology.
- Nanotechnology can also help in waste management and environmental remediation.

Nanotechnology for Green Manufacture

Nanotechnology can change the manufacturing process in two ways:

- Using nanotechnology for controlled, effective manufacture will reduce waste drastically.
- Using nanomaterials as catalysts will ensure better efficiency in present manufacturing processes and will reduce or eliminate the use of toxic materials and generation of undesirable by-products and effluents.

Research has included using nanotechnology to enhance industrial processes and materials, developing new industrial and chemical procedures, developing materials that will act as a replacement for hazardous processes and constituents causing reductions in materials, energy and waste generation.

Listed below are potential examples of nanotechnology research that may cause elimination or reduction of pollutants of concern:

- Atomic level synthesis of novel and improved catalysts for industrial processes.
- Addition of information to molecules to build new molecules.
- Self-assembly of molecules as the basis of new materials and chemicals.
- Building molecules in microscale reactors.

Nanomaterials for Water Treatment

The availability of safe, clean drinking water is a challenge being faced worldwide, especially in developing countries. Water purification methods are chemical intensive and are costly as well as harmful to the environment; they cannot be adapted easily to the non-industrialized world. Nanomaterial-based adsorbents, catalysts and technologies can create eco-friendly solutions for water treatment.

Nanomaterials can be used for detecting and sensing pollutants, treating and remediating contaminants and preventing pollution. Nanomaterials are also used to improve membrane separation processes, resulting in better selectivity and lower costs.

Studies indicate that nanotechnology-based water treatment techniques will be more economical, more durable, and more efficient than the present, traditional ones. Technologies that are available commercially or are under development make use of nanomaterials in meshes, filters, clays, ceramics and adsorbents, catalysts and zeolites.

An example of the contaminant filter technology is carbon nanotube (CNT) membranes. They have a high permeability, increased surface area and good thermal and mechanical stability. They can remove water components such as bacteria, stirred up sediments, viruses and organic contaminants. They will require lesser maintenance when compared to osmosis membranes.

Nanotechnology for Renewable Energy

A research area that is being extensively studied is the ability to harness green nanotechnology to design solar cells. Some novel solar cell materials which have been suggested include nanoparticles such as titanium dioxide, cadmium telluride, quantum dots and silver with a polymer that can absorb solar energy. According to the researchers, the cost of production of these solar cells will be much less than those commercially available now. However, the efficiency of these cells are not very high, hence researchers are working to obtain the desired efficiency.

Other techniques include depositing nanocrystals, nanoparticles suspended in quantum dots, nanoparticle ink, silver cells and nanowires and producing highly stable laminate layers for solar cell protection.

A key challenge in renewable energy technology is that power generation is quite inconsistent and has to be on a largescale. Efficient batteries are required to store extra energy produced during windy or sunny periods, so that it can be used during lulls to ensure there is always a steady supply. It is anticipated that the performance of energy storage technologies can be improved using nanotechnology to enable long-life, high capacity energy storage devices for renewable energy applications.

Nanotechnology for Environmental Remediation and Waste Management

There are several proposed remediation techniques that use nanotechnology. For example, solar photocatalysis using titanium dioxide nanoparticles can degrade pollutants like volatile organic compounds and nitrous oxides and consequently has been used in cement and commercial plants. Paints enriched with nano titanium dioxide, replacing organic biocides, are used to keep the surfaces of buildings clean.

Soil remediation of abandoned and old military and industrial sites has become a great challenge in industrialized nations. Trials have been set-up using nano zero-valent iron and iron oxides and the results have been quite promising. However, there are environmental concerns regarding the impact of nanoparticles on the soil.

Role of Nanotechnology in Chemical Substitution

In the area of coatings and adhesives, nanotechnology products are proposed as potential substitutes for harmful chemicals such as highly toxic chemicals and heavy metals. It is proposed that nanoparticles of silicon dioxide, titanium dioxide, magnesium oxide or zinc oxide could replace chemical flame retardants such as bromine, which is considered to have high toxicity.

Environmental Concerns with Nanotechnology

One major issue with nanotechnology is that it is difficult to keep track of nanoparticles once they are released into the environment. Therefore, the impact of nanoparticles on ecosystems cannot be determined easily. Long-term exposure to low, highly dispersed doses of nanoparticles may cause a considerable impact - however, this kind of exposure is very tough to model in trials. It is essential that consumers are well educated, to make sure nano-waste streams are under control or at least well understood.

More research needs to be done with regards to structure-function relationships and in correlating chemistry and surface area to toxicity and functionality. Risk assessments must be performed on nanomaterials presenting a real exposure risk during use or manufacture. These findings must consider the toxicological hazard, the biological and environmental fate, the probability of being exposed, persistence, transport, transforming into the completed product, and recycling.

NANOTECHNOLOGY IN SUSTAINABLE AGRICULTURE

Nanotechnology monitors a leading agricultural controlling process, especially by its miniature dimension. Additionally, many potential benefits such as enhancement of food quality and safety, reduction of agricultural inputs, enrichment of absorbing nanoscale nutrients from the soil, etc. allow the application of nanotechnology to be resonant encumbrance. Agriculture, food, and natural resources are a part of those challenges like sustainability, susceptibility, human health, and healthy life. The ambition of nanomaterials in agriculture is to reduce the amount of spread chemicals, minimize nutrient losses in fertilization and increased yield through pest and nutrient management. Nanotechnology has the prospective to improve the agriculture and food industry with novel nanotools for the controlling of rapid disease diagnostic, enhancing the capacity of plants to absorb nutrients among others. The significant interests of using nanotechnology in agriculture includes specific applications like nanofertilizers and nanopesticides to trail products and nutrients levels to increase the productivity without decontamination of soils, waters, and protection against several insect pest and microbial diseases. Nanotechnology may act as sensors for monitoring soil quality of agricultural field and thus it maintain the health of agricultural plants.

Agriculture is always most important and stable sector because it produces and provides raw materials for food and feed industries. The limit of natural resources (production land, water, soil, etc.) and the growth of population in the world claim the agricultural development to be economically further, viable, environmentally and efficiently. This alteration will be the vital for achieving many factors in the recent year. Agricultural nutrient balances are differed noticeably with economic growth, and especially from this surmise, the development of the soil fertility is very much significant in developing countries.

The development of agriculture is compulsory phenomena for the purge of poverty and hunger which must be getting rid of from the present situation. Therefore, we should have to take one bold step for agriculture development. In this world mainstream of peoples are below poverty level which are being scatted in the rural area where agriculture enlargement has not so being effective.

Nowadays, the most vital obsession is to create flanked by, agriculture poverty and nutritional process getting food. Therefore, new technology should have to adopt that decidedly focuses on getting better agricultural production. Recently, food and nutritional security are fully embedded in the novel knowledge. The agriculture development also depends on the social inclusion, health, climate changes, energy, ecosystem processes, natural resources, good supremacy, etc., must also be documented in specific target oriented goals. Therefore, sustainable agricultural strengthening the practical opportunity to get rid of poverty and hunger of the people. The agriculture on the road to recovery, thus the environmental performance is required and at the same time participation of food chain ecosystems are required in relation to agricultural food production.

No doubt that the sustainable growth of agriculture totally depends on the new and innovative techniques like nanotechnology. Naturally, it haunts us to know what is this important technology? If we like to go in the year 1959 Feynman's lecture on "Plenty of room at the bottom," from this very day, the nanoprocess is in underway. Later on Professor Norio Tanaguchi proposed the actual term of nanotechnology. Afterward, nanotechnology develops more vivid way, as because, more recent instruments develops to consider or isolate nanomaterials in accurate way. Additionally, the number of publications related to the term of "nano" was also grown exponentially. Figure demonstrates the number of documents on scopus.com with the search term of "Nano and (Food and Agriculture)." In 2016, about 14,000 documents with nanotechnology in food or agriculture were listed meaning high activities of this field. Also about 2707 patents matched this criteria are found in world patent database. The world market size of nanotechnology in 2002 was about US$ 110.6 billion and predicted to grow to US$ 891.1 billion in 2015 according to analysis of Helmut Kaiser Consultancy. The developments of nanotechnology in materials and electronics have higher dynamics than other applications. Recently, food and agriculture also require high amount of nanomaterials especially in packaging. The NAFTA region shares the biggest slice from the market size, but Europe and Asia especially China, Japan, and India also come up very dynamically.

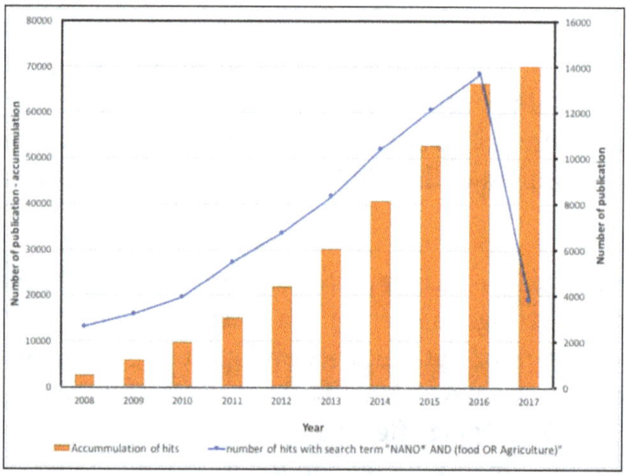

Number of documents on scopus.com with the search term "Nano and (food or agriculture). The hits were grouped annually.

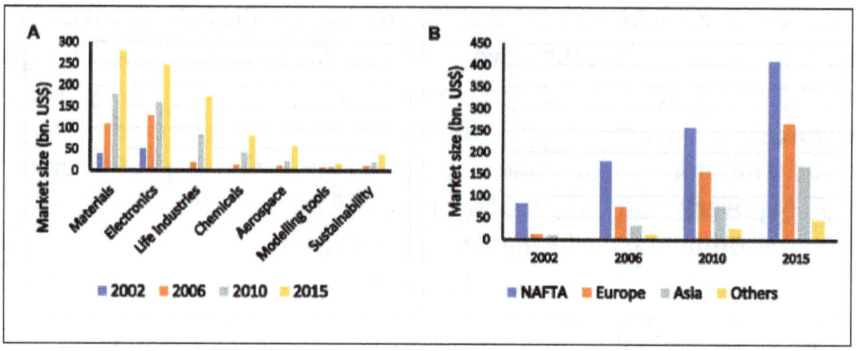

World market of nanotechnology by applications (A) and regions (B).

It is well known that one billionth of a meter is one nanometer (nm). Why this nano will change its property? It is due to alteration in the atoms and develops a magnetic power. It can be projected that the smaller size of nanomaterial possesses larger surface area and exhibits more active. With this course of action nanotechnology is knocking the doors of perception. The magnetic property of polymer develops due to tellurium atoms; antimony-bismuth; and sulfur atoms. Moreover, it has been observed that when the atoms of do pant and atoms of europium interact together, and then the entire molecules carry out the magnetic property. Thus the alter property of nanomaterials is related with more reactive in the most sectors including in biological process. This ultimate technology possesses several unique electronic association, plasmonic and optical properties which are related with the quantum confinement effects, the alteration of the electronic energy levels may appear due to the surface area in relation to volume ratio. In the present century, there is a big demand for fast, reliable, and low-cost systems for the detection, monitoring, and diagnosis for biological host molecules in agricultural sectors. The application of chemically synthesize nanomaterials now a days considered as toxic in the nature, in attract of this, nanomaterials may synthesis from plant system and it considered as green nanotechnology. Green nanotechnology is a safe process, energy efficient, reduces waste and lessens greenhouse gas emissions. Use of renewable materials in production of such products is beneficial, thus these processes have low influence on the environment. Nanomaterials are eco-environmentally sustainable and significant advances have been made in the field of green nanotechnology. In the present decade, it is more shift toward the green nano in a faster rate for implementation its functions. Still it is not clear how the environmental sustainability of green nanotechnology will be achieved in future? These risks must be mitigated in advancing green nanotechnology solutions.

In modern agriculture, sustainable production and efficiency are unimaginable without the use of agrochemicals such as pesticides, fertilizers, etc. However, every agrochemical has some potential issues including contamination of water or residues on food products that threat the human being and environmental health, thus the precise management and control of inputs could allow to reduce these risks. The development of the high-tech agricultural system with use of engineered smart nanotools could be excellent strategy to make a revolution in agricultural practices, and thus reduce and/or eliminate the influence of modern agriculture on the environment as well as to enhance both the quality and quantity of yields.

The development of biosensors is also a good field for exploitation of many strengths of nanotechnology, thus nanotechnology is there and plays an essential role. Due to special properties of nanomaterials, on one hand, the sensitivity and performance of biosensors could be improved significantly in their applications; on another hand, many new signal transduction technologies are let to be introduced in biosensors. Additionally, use of nanomaterials let to miniaturize many (bio)sensors to small and compact/smart devices such as nanosensors and other nanosystems that are very important in biochemical analysis. It also helps to detect the mycotoxins present in several foods and their functions are very rapid.

Nanotechnology and Agricultural Sustainable Development

The nanotechnology can takes an important part in the productivity through control of nutrients as well as it can also participate in the monitoring of water quality and pesticides for sustainable development of agriculture. Nanomaterials have such diverse assets and activities that it is impossible to deliver a general assessment of their health and environmental risks. Properties (other than size) of NPs have the influence on toxicity include chemical composition, shape, surface structure, surface charge, behavior, extent of particle aggregation (clumping) or disaggregation, etc. may associate with engineered NPs. For this reason even nanomaterials of the same chemical composition that have different sizes or shapes can exhibit their different toxicity. The implication of the nanotechnology research in the agricultural sector is become to be necessary even key factor for the sustainable developments. In the agri-food areas pertinent applications of nanotubes, fullerenes, biosensors, controlled delivery systems, nanofiltration, etc. were observed. This technology was proved to be as good in resources management of agricultural field, drug delivery mechanisms in plants and helps to maintain the soils fertility. Moreover, it is being also evaluated steadily in the use of biomass and agricultural waste as well as in food processing and food packaging system as well as risk assessment. Recently, nanosensors are widely applied in the agriculture due to their strengths and fast for environmental monitoring of contamination in the soils and in the water. Several sensors based on nano-detection technology such as viz. biosensors, electrochemical sensors, optical sensors, and devices will be the main instruments for detecting the heavy metals in trace range.

Nanomaterials not only directly catalyze degradation of waste and toxic materials but it also aids improve the efficiency of microorganisms in degradation of waste and toxic materials. Bioremediation uses living organisms to break down or remove toxins and harmful substances from agricultural soil and water. In particular, some other terms are also generally used such as bioremediation (beneficial microbes), phytoremediation (plants), and mycoremediation (fungi and mushrooms). Thus, with the bioremediation the heavy metals can be removed from soil and water environmentally and efficiently by microorganisms. Therefore, the agricultural bioremediation helps in sustainable remediation technologies to resolve and restore the natural situation of the soil. It is an interesting phenomena in considering the nano–nano interaction to remove the toxic component of the agricultural soil and make it sustainable.

Nanofertilizers

In the recent decade nanofertilizers are freely available in the market, but particularly the agricultural fertilizers are still not shaped by the major chemical companies. Nanofertilizers may contain nano zinc, silica, iron and titanium dioxide, ZnCdSe/ZnS core shell QDs, InP/ZnS core shell QDs, Mn/ZnSe QDs, gold nanorods, core shell QDs, etc. as well as should endorse control release and improve the its quality. Studies of the uptake, biological fate and toxicity of several metal oxide NPs, *viz.* Al_2O_3, TiO_2, CeO_2,

FeO, and ZnONPs were carried out intensively in the present decade for agricultural production. The deficiency of zinc has been documented as one of the main problems in limiting agricultural productivity in the alkaline nature of soils.

Metal oxide NPs are radiolabeled by direct proton bombardment or enriched during synthesis with ^{18}O to generate ^{18}F. Size, degree of aggregation and zeta potential of the metal oxide NPs are studied in the presence of proteins and cell media. Moreover, NP uptake and intracellular fate are followed by ion beam microscopy, transmission electron microscopy, Raman chemical imaging spectroscopy, and confocal laser scanning microscopy. In the future, sustainable bio-based economy that uses eco-efficient bio-processes and renewable bio-resource, will continue decrease and substitute the harmful materials in established applications, and thus it will play a major role (the key strategic challenge) in the development of the technologies desired to address to 21st century. Accumulation of knowledge in fields of ecology, biology, biodiversity, material science, biotechnology, and engineering opens possibilities to increase biomass productivity as well as to utilize biomass and organic wastes at a highly efficient.

References

- Environmental-nanotechnology: understandingnano.com, Retrieved 13 July, 2020
- Crane, R. A.; T. B. Scott (2012-04-15). "Nanoscale zero-valent iron: Future prospects for an emerging water treatment technology". Journal of Hazardous Materials. Nanotechnologies for the Treatment of Water, Air and Soil. 211–212: 112–125. doi:10.1016/j.jhazmat.2011.11.073. ISSN 0304-3894. PMID 22305041
- Roy, Yagnaseni; Warsinger, David M.; Lienhard, John H. (2017). "Effect of temperature on ion transport in nanofiltration membranes: Diffusion, convection and electromigration". Desalination. 420: 241–257. doi:10.1016/j.desal.2017.07.020. hdl:1721.1/110933. ISSN 0011-9164
- Voisin, H; Bergström, L; Liu, P; Mathew, AP (2017). "Nanocellulose-Based Materials for Water Purification". Nanomaterials. 7 (3): 57. doi:10.3390/nano7030057. PMC 5388159. PMID 28336891
- Turkevich, Leonid A.; Fernback, Joseph; Dastidar, Ashok G.; Osterberg, Paul (2016-05-01). "Potential explosion hazard of carbonaceous nanoparticles: screening of allotropes". Combustion and Flame. 167: 218–227. doi:10.1016/j.combustflame.2016.02.010. PMC 4959120
- Buzea, Cristina; Pacheco, Ivan I.; Robbie, Kevin (December 2007). "Nanomaterials and nanoparticles: sources and toxicity". Biointerphases. 2 (4): MR17–71. arXiv:0801.3280. doi:10.1116/1.2815690. PMID 20419892

Permissions

All chapters in this book are published with permission under the Creative Commons Attribution Share Alike License or equivalent. Every chapter published in this book has been scrutinized by our experts. Their significance has been extensively debated. The topics covered herein carry significant information for a comprehensive understanding. They may even be implemented as practical applications or may be referred to as a beginning point for further studies.

We would like to thank the editorial team for lending their expertise to make the book truly unique. They have played a crucial role in the development of this book. Without their invaluable contributions this book wouldn't have been possible. They have made vital efforts to compile up to date information on the varied aspects of this subject to make this book a valuable addition to the collection of many professionals and students.

This book was conceptualized with the vision of imparting up-to-date and integrated information in this field. To ensure the same, a matchless editorial board was set up. Every individual on the board went through rigorous rounds of assessment to prove their worth. After which they invested a large part of their time researching and compiling the most relevant data for our readers.

The editorial board has been involved in producing this book since its inception. They have spent rigorous hours researching and exploring the diverse topics which have resulted in the successful publishing of this book. They have passed on their knowledge of decades through this book. To expedite this challenging task, the publisher supported the team at every step. A small team of assistant editors was also appointed to further simplify the editing procedure and attain best results for the readers.

Apart from the editorial board, the designing team has also invested a significant amount of their time in understanding the subject and creating the most relevant covers. They scrutinized every image to scout for the most suitable representation of the subject and create an appropriate cover for the book.

The publishing team has been an ardent support to the editorial, designing and production team. Their endless efforts to recruit the best for this project, has resulted in the accomplishment of this book. They are a veteran in the field of academics and their pool of knowledge is as vast as their experience in printing. Their expertise and guidance has proved useful at every step. Their uncompromising quality standards have made this book an exceptional effort. Their encouragement from time to time has been an inspiration for everyone.

The publisher and the editorial board hope that this book will prove to be a valuable piece of knowledge for students, practitioners and scholars across the globe.

Index

www.ingramcontent.com/pod-product-compliance
Lightning Source LLC
LaVergne TN
LVHW080635210926
848485LV00003B/267